Conflict

Conflict

How Soldiers Make Impossible Decisions

Neil D. Shortland, Laurence J. Alison,
and Joseph M. Moran

UNIVERSITY PRESS

Oxford University Press is a department of the University of Oxford. It furthers
the University's objective of excellence in research, scholarship, and education
by publishing worldwide. Oxford is a registered trade mark of Oxford University
Press in the UK and certain other countries.

Published in the United States of America by Oxford University Press
198 Madison Avenue, New York, NY 10016, United States of America.

© Oxford University Press 2019

All rights reserved. No part of this publication may be reproduced, stored in
a retrieval system, or transmitted, in any form or by any means, without the
prior permission in writing of Oxford University Press, or as expressly permitted
by law, by license, or under terms agreed with the appropriate reproduction
rights organization. Inquiries concerning reproduction outside the scope of the
above should be sent to the Rights Department, Oxford University Press, at the
address above.

You must not circulate this work in any other form
and you must impose this same condition on any acquirer.

Library of Congress Cataloging-in-Publication Data
Names: Shortland, Neil D., author. | Alison, Laurence J., author. |
Moran, Joseph M., author.
Title: Conflict : how soldiers make impossible decisions / Neil D. Shortland,
Laurence J. Alison, Joseph M. Moran.
Description: New York : Oxford University Press, [2019] |
Includes bibliographical references.
Identifiers: LCCN 2018022509 (print) | LCCN 2018047021 (ebook) |
ISBN 9780190623456 (eBook) | ISBN 9780190623449 |
ISBN 9780190623449 (hardcover)
Subjects: LCSH: War—Decision making. | Combat—Psychological aspects. |
Soldiers—Psychology. | Psychology, Military.
Classification: LCC U22.3 (ebook) | LCC U22.3.S47 2019 (print) |
DDC 355.001/9—dc23
LC record available at https://lccn.loc.gov/2018022509

This book was made possible by the willingness of soldiers to share experiences with us in interviews. It was an honor and a privilege to hear their accounts and have the opportunity to learn about how they made incredibly hard decisions in incredibly hard situations. It is with great pleasure and gratitude that we dedicate this book the all members of the Armed Forces. Twenty-five percent of the authors' personal profits from this book will be donated to charities in the United States and the United Kingdom dedicated to supporting the soldiers (and their families) who have served in the Armed Forces.

CONTENTS

Foreword ix
Col. Tim Collins
Preface xv
Acknowledgments xxi

1. Decisional Conflict: From Best to Least-Worst 1
2. Military Decision-Making: Doctrine, Rationality, and Field-Based Approaches 15
3. The Science of Selecting Least-Worst Options 34
4. Situation Awareness 53
5. Formulating Plans 76
6. Executing Plans 101
7. Team Learning 120
8. Least-Worst Decision-Making "in Extremis" 138
9. Thoughts That Haunt 160
10. How Do Soldiers Do What They Do and What Can We Learn from It? 175

References 191
Index 211

FOREWORD

I served for 23 years in the regular British Army and before that several years in the army reserve as well as the Ulster Defence Regiment, a locally recruited militia during the troubles in Northern Ireland. I had joined in the days after my 17th birthday and had been a schoolboy by day and soldier at night until my actual age was discovered and I was sent to the Territorial Army. Since leaving the Army, I have served as a battlefield contractor leading training teams in Iraq and Afghanistan alongside the US Military and the US Marine Corps in particular.

In that capacity, I have seen the evolution of military decision-making from a vicious internal conflict in Northern Ireland—one described as the "corporal's war" in the early years because the corporals were the main decision-makers and shouldered most of the responsibility for the decisions—through the Cold War in Europe, during which decisions were subsets of already well-rehearsed plans. I went on to serve with the British Army's Special Forces, the Special Air Service, where I was a troop commander on numerous overseas operations and promoted to Operations Officer of 22 SAS and later Staff Officer Grade 1 (SO1 Ops) as a Lieutenant Colonel in charge of worldwide operations for UK Special Forces (UKSF). This was a role that required me to sit in on meetings in the Cabinet Office Briefing Room A (COBRA), the UK's military emergency decision-making forum. I have also commanded a battalion on Internal Security (IS) operations in Ulster and led my men in general war in the liberation—or invasion (as you please)—of Iraq. As a civilian, I have found that leading civilians in conflict zones is trickier than leading soldiers because they are civilians and not soldiers. *Civilians don't have to do what they are told*. I have briefed senior officers on many operations and as SO1 Ops helped them make decisions about the arrest of persons indicted for war crimes in the former Yugoslavia (PIFWCs). I have benefitted from many of the decisions I have made or recommended, because success brings advancement and I have also suffered, from the consequences of my decisions—I was accused of war crimes for acting to save the lives of my soldiers and their informants in Iraq, and although cleared of any wrongdoing, per se, my career was finished.

And yet, as if it was preordained, in civilian life I found myself more influential in many ways than ever in military decision-making. The company I lead works at the highest level of government in the countries in which we work, and I must admit to being hugely humbled by the level of access the US military extends to me and amazed by their response to advice.

This is enhanced by the fact that the US military is, in many ways, years ahead of the UK military, and the robustness of US commanders, contrasted to UK commanders, is stark. Take, for example, the Iraq War. The multi-pound Chilcot investigation into the war found that "the UK military role in Iraq ended a very long way from success" and that

> by 2007 militia dominance in Basra, which UK military commanders were unable to challenge, led to the UK exchanging detainee releases for an end to the targeting of its forces.
>
> It was humiliating that the UK reached a position in which an agreement with a militia group which had been actively targeting UK forces was considered the best option available.

Although there has been no equivalent US inquiry, I have seen the US commanders up close and sat in on decision-making. They are the opposite of what is described by Chilcot. I recall when the bodies of Marines were being routinely defiled and mutilated in Anbar Province, Iraq, the commanding general suddenly found the necessity of retaining insurgent bodies for forensic and evidential reasons. In a country in which the religion required a fast burial, long lines of bloated black bodies festering in the sun and in plain public view behind barbed wire fences soon ended the mutilation of the Marine bodies, and a clear understanding between the combatants was established. Would a British general have done the same for his men's dignity? I very much doubt that.

In my many guises, I have been fortunate to see the military as well as the politicians who dictate policy in action. In Northern Ireland, the politicians eventually assumed a pretty front-line role as the corporal's war became the superintendent's war, with police primacy, and then became a politician's war as the players on both sides played an increasingly political role. This was the period that led to the peace process. This was, by the time I was a senior UKSF staff officer, acute, and the activities of the SAS were strategic with the "green army" as we called the regular troops, simply holding ground, shaping operations, and protecting the police.

What I then saw clearly a couple of years later, as the commanding officer of a green army regiment (the Royal Irish Regiment, an air assault battalion) on a 6-month emergency tour, in contrast to my memories of the late 1970s, the change was stark. What also stuck me, then and now, was the extent to which many more senior officers did not really understand what was going on in Northern Ireland at the time or since.

Yet today when my company mentors the Afghan police Special Branch, it is clear that the depth of understanding of US officers, who have never set foot in Ireland, North or South, instinctively understand the power of police intelligence during an insurgency and the need to act on it but to also protect the source in a manner far in advance of their UK counterparts.

And this is the key point. Acting in an intelligence-led way is by far the most effective means of prosecuting a counter-insurgency war and the most risky. Suddenly, the life of an informant or an agent can become more important than the lives of many servicemen only because the power to make the right decision and ultimately defeat the enemy and preserve life is so profoundly affected by that information. This is the long war and one that is difficult to get right. The US Armed Forces, it must be said, are the masters in this regard above any other coalition partners.

Much of the serious damage done to the mindset of the UK Army decision-making capabilities during the past 14 years and not seen in the United States (yet) is due to the "you're on your own culture." Many field commanders are reduced to the level of scout masters—that is, the fear of litigation over their charges or responsibility for the actions of their charges is their only concern (a case in Chapter 4 highlights this phenomenon). Getting the mission done has no standing now at all. Apparently, it has been trained or frightened out of them. A new phenomena exists in which, in a bind, the right thing to do is nothing—known as "courageous restraint," coined during the McChrystal era. Essentially "do nothing." It was designed to put some restraint on the more robust US commanders, but by the late 2000s, the effect it had on UK commanders, with any aggressive spirit already trained out of them, was stark. Senior UK commanders thrive and live by this code. The lives of our young servicemen and women are expendable against the danger of accusations of excess or even actual war crimes allegations. Accusations of that nature are the individual's problem; the UK military will not stand by the individual at all, from top to bottom. In fact, members of the military are encouraged to cooperate and exaggerate in order to protect themselves. Decision-making, or in contrast obfuscation or refusal to do so, has thus become the main vehicle for self-preservation.

The effect has become such that making a decision that would cause the loss of one or a few soldiers is now an understandable outcome of operations, and the full might of the Ministry of Defence will help back one up—even bend the truth to cover moral cowardice or actual cowardice. It starts at the top.

A state of affairs exposed in Chilcot found that the young British servicemen in Iraq, even under severe attack by improvised explosive devices and snipers, were told to act as if this were normal. The message was, "Behave as if it was a benign Northern

Ireland-type situation in order to protect the PM's reputation and statements to Parliament." Commanders were reporting "All good here—Snatch Land Rovers are just fine" even as they knew they were not but realizing that the first to say "We're getting slaughtered, we need heavier armor" would be immediately sacked (and was). Once more, doing nothing was the highest priority.

It is exactly the same phenomena that occurred with the German generals who refused to tell Hitler in his forward HQ the true extent of Russian advances for fear of his reaction. So, for instance, using the huge Russian numbers against them by drawing the Russians into a trap and then snapping them off while overextended and undersupplied never happened because that maneuver required a withdrawal and all but Manstein refused to order this. Hitler demanded, mainly for propaganda and reputational reasons, that every inch of ground was contested, allowing the Russians to build huge strength at sectors of their choosing. Manstein, who could have avoided the situation, was sacked.

How can this be? In an ironic twist, such timidity almost always has its roots in success. In 1805, Nelson's fleet smashed the combined French and Spanish fleets at Trafalgar. Success was in large part due to the speed of decision-making by the Royal Navy based on mission command and delegation of authority for decision-making at the tactical level, in contrast to the centralized flag-based control favored by Admiral Villeneuve. After the victory and the death of Nelson, the Lords of the Admiralty sought to preserve, as if in aspic, that successful leadership and fighting spirit for future generations. Rather than seek to understand such decision-making and how it was developed, they ordered it to be copied and taught—as if that could be an alternative to a natural development. Now, what is learned by instinct cannot be taught; it must be learned by example and nurtured by delegation. What the lords of the Admiralty had either not known or forgotten was that the commanders of ships were a breed that was evolved of experience of privations, long periods at sea, and often being outnumbered, outgunned, and hunted. They were the celebrated "rat catchers." Nelson, Hood, and Collingwood thought like a rat to catch a rat.

They were replaced by the "regulators." The regulators wrote books, measured things, weighed things, but did not understand anything. To get on in the Navy for the next century, one had to read their books and pass exams based on these books. Decision-making was now a formula, not an art form borne of experience, intuition, and training—the issues of which are discussed in Chapter 2. So the potential for poor decision-making and thus errors steadily increased. By 1916, when the British Grand Fleet met the German High Seas Fleet at Jutland, the job was done. British decision-making capabilities—despite help from access to German naval codes—was fatally undermined by inertia and adherence to decision-making rules, on the one hand, and by a blatant disregard for safety measures, on the other hand, in order to achieve the execution of these decisions. In terms of technology, the playing field was fairly balanced. Both fleets had "dreadnought" battleships with guns so powerful they could destroy lesser ships far before the dreadnought ships could be engaged, turbo diesel engines to provide speed, and armor that ordinary ships could not penetrate. They also had safety devices, such as water- and flash-proof compartments, for protection against other dreadnoughts.

But the British leadership, Jellicoe and Beattie, were brought up in the regulator's care, and they believed that aggressive spirit was about getting in close and firing guns quickly—for that is what Nelson did—as opposed to leveraging technology and intelligence to make sound decisions. In doing so, fatal errors developed, including a recklessly aggressive dash. This resulted in the common practice of overriding the safety procedures by propping the waterproof and flashproof hatches open with the cordite charge bags that fired the huge shells, which would in turn be raised by lifts to the turrets. "Don't starve the guns" was the cry, and to fire the guns quickly and aggressively was the acme of naval power. In doing so, the captains had turned their ships, literally, into powder kegs.

The German Navy did not have the benefit of 100 years of regulators. Admiral Scheer led by mission command. The Germans recognized that with the available dreadnought technology and timely decision-making, the acme of naval power was to sink the enemy's ships. To do so, they employed all of the available defensive and safety measures when being fired upon and, conversely, when firing, did so very precisely and accurately with the aim of sinking the enemy. On the day they sank twice as many Royal Navy ships as they lost, they nonetheless had to return to Keil—never to emerge again because they were overwhelmed by numbers. The point is that their decisions were based on the technology and needs—not an overblown myth or pipe dream.

Today, the flawed decision-making that dogs UK commanders in Iraq and Afghanistan comes of a desire to fight the last war they won, the counter-insurgency in Northern Ireland. The problem is that they did not win it. The Royal Ulster Constabulary (RUC) Special Branch ultimately won it with the help of the Army. The RUC penetrated the Republican Movement so completely that it had to give in to the British politicians. It was support to the Civil Power.

That is not to say there had not been a military war—there had been at the beginning in the 1970s and early 1980s a corporal's war, which was vital for setting the conditions for ultimate victory by allowing the police to enter the stage, and military primacy thus ended in the mid-1980s. The problem was that no one who had been in a command position in Northern Ireland at a sufficient level was still serving by the time Afghanistan and Iraq were invaded. The lessons were lost. Decision-making in the theater was based on a fallacy of what they imagined had been the norm in Northern Ireland.

The US military, in contrast, unhampered by this experience and despite the lectures on Northern Ireland tactics from British Commanders, fought the corporal's war successfully. Unfortunately, they left Iraq—for political expedience—before the rule of law phase could begin, and sectarian tensions and deliberate sabotage by Iran have left Iraq a husk. Had the US military persisted, success was achievable and the defeat of the insurgency inevitable. There is still hope in Afghanistan, however, but another surge may be required.

But here is the point. The US servicemen and women who served in Iraq fought the corporal's war. Their commanders rapidly understood that decision-making based on existing drills and standard operating procedures was only part of what was needed. Things would happen out on the ground that demanded instant fresh thinking and courageous, bold decisions. The decisions the corporals made every day made the difference in a war of attrition. But, as highlighted in Chapter 9, they live with the consequences of those decisions, which affect them much more than the decisions made at the operational or strategic level. This book examines the subject and provides a basis for those who follow to understand.

Colonel Timothy Collins, OBE

PREFACE

I would rather go down this street and get in a firefight than go back and get blown up by an IED. This thought was resounding, and I would not forget it, I would rather get in a firefight than face an IED.

This book is about difficult decisions. It explores the mental conflict that people experience when faced with choices for which all outcomes appear horrendous. It draws on interviews that we conducted with real soldiers[1] in real situations and focuses specifically on the impossible choices that they had to make—choices that were vivid and sometimes upsetting for them and yet were part of the experience of war and which have stayed with them throughout their lives. The book is about the internal conflict that they experienced when forced to make "least-worst" decisions.

In this book, we discuss the methods we have used (as psychologists) to understand and measure this conflict and when and why it emerges. The book describes the processes that make choice selection so difficult; the psychology of decisional conflict; and the immediate, short-term, and longer term behavioral consequences of experiencing this conflict. Special attention is directed to the concept of "decision inertia"—one of several kinds of "failures to act" in which decision-makers are unable to calculate and/or commit to a least-worst course of action[2] (normally because people are so averse to a bad outcome—even if it is the least bad of two or more awful options). Emergency services and law enforcement personnel seem especially prone to the inability to tolerate any bad outcome—the military far less so. In later chapters, we sketch our analysis of *why* we think soldiers are more tolerant of a least-worst option, and thus less prone to decision inertia. Finally, by reference to many real cases of decisional conflict, as well as by asking readers to consider their own responses to different types of conflict scenarios, we hope to illustrate the fundamental theoretical

---

[1] Although the term "soldier" typically denotes members of the Army, in using this term we are referring to members of all branches of the Armed Forces.

[2] We use the term "course of action" throughout this book because it is a relatively well-used piece of military parlance. For the nonmilitary reader, course of action (later shortened to CoA) is, simply, a potential action that can be taken to (it is hoped) bring about a desired outcome. In this sense, it is not dissimilar to a "choice" as outlined more generally by psychologists who study decision-making (see Chapter 1).

and practical importance of this concept. And, in no small part, we want to help readers understand what it is really like to be faced with options that no sane individual would willingly choose.

The quote at the beginning of this chapter was recounted to one of the authors (N. Shortland) in one of the first interviews conducted for this book, and it describes the dilemma of a convoy leader in Iraq. While providing security for a convoy running between Baghdad and a forward operating base, he found that his planned routes (route A, preferred; route B, not ideal but workable) had been blocked, randomly, as often occurs in Iraq. A quick decision was needed because they were fast approaching the rotary (a "roundabout" for UK readers). Option 1 was to take route "C," a new, unplanned route that would lead directly through a "troubled" part of town—meaning that it was a known hot spot for insurgent fighters, and most times coalition forces had navigated this part of town, they had come under fire. Option 2 was to turn around, backtrack, and then go around the center of the city (rather than through it via route C, which was both shorter and faster, getting them and their cargo back to base sooner). A further wrinkle was that the convoy leader knew insurgent fighters laid improvised explosive devices (IEDs) on common coalition routes, meaning there was a high chance the route they had just traveled (and were now considering backtracking through) would be laced with IEDs. In challenging physiological and psychological conditions, he was forced to make a decision that had major consequences; there was no "right" answer (at least not an answer that would be apparent until *after* the decision was made), high uncertainty, and little time. He was unlikely to have made a decision like this in training, nor had he faced one during his previous tour of Afghanistan.

So, how did he make his decision? And how can psychologists better understand the processes people use to make impossible decisions in challenging environments?

These were the questions in our minds when we began examining military decision-making. We were specifically interested in least-worst decisions, which we have observed extensively in our research on police and emergency service decision-making that we have been conducting during the past few decades. Strangely, however, very little work has explored least-worst decisions in a military context. We set out on a journey to understand what least-worst decisions in the time-limited, information-starved, inveterately ambiguous heat of battle mean for decision-makers and how they perform the calculus needed to accept a bad outcome. We also wanted to understand the future consequences of having to choose between two or more awful outcomes for which "all routes look bad": What does this mean for one's future mental state? The existing research has often studied how soldiers are trained to make decisions, and it very rarely collects data from the points at which real decisions are made and the *types* of decisions made when it counts: at war.

THE INTERVIEWS AND THE INTERVIEW METHOD

We conducted a series of critical decision method (CDM) interviews (Crandall, Klein, & Hoffman, 2006) with currently serving and retired members of the US Armed Forces. We interviewed servicemen and women across all branches (Army, Navy, Marine Corps, Air Force, and Coast Guard) and across a diverse range of ranks. CDM interviews involve four "sweeps," which use different types of probes and perspectives in an attempt to facilitate the quality of recall (Crandall et al., 2006). The first sweep results in the selection of an incident that matches the requirements of the research and the goals for data collection. The second sweep involves the participant recalling the event from start to finish. The third sweep involves the interviewer using a series of cues to investigate the participant's experience of the event. This sweep seeks to go beyond the timeline and understand the participant's perceptions, expectations, goals, and uncertainties during the incident. It is in this sweep that we gain the greatest insight into the participant's decision-making. The final sweep involves questioning all the "what ifs" of an event in order to identify any factors (within the environment or related to the decision-maker) that *could*, if different, have resulted in a different outcome (Crandall et al., 2006, pp. 69–83).

Our CDM interviews started with the following statement:

> I am going to be asking you in a moment to spend some time thinking about a decision that you had to make, while in the Armed Forces, in which you had to choose between one or more options and in which you spent a lot of time thinking about all the possible outcomes.

Perhaps, as the reader, you would like to take some time to think of a situation that matches this brief. Maybe you had to make a major life-changing decision: Changing your career? Making a large purchase (a house, etc.)? Or perhaps ending (or starting) a serious relationship? Once you have thought of a candidate decision, perhaps you can think about the amount of time it took you to mull it over. Did you avoid making the decision? Did you spend a lot of time thinking about it? Did you come to a conclusion (i.e., what you *want* to do) but then not take the required steps to put this preference into action? If you answered "yes" to any of these questions, you experienced what we call decision inertia. Contrary to the decades (even centuries) of research aimed at understanding how people make decisions and what makes a "good" decision-maker (and in a military setting, how individuals can be selected, trained, and developed to be good decision-makers for leadership positions), very little psychological research has focused on the process of decision inertia (and it is indeed a process). Perhaps there is a view that incorrect decisions are costlier than indecision (and hence warrant more pressing research attention). We do not agree. At the personal and societal level, indecision has significant consequences. Consider Syria, in the first 3 years of the war, there was clear inertia about intervention, and it is arguable (albeit still counterfactual) that *any* decision to intervene in *any* way could have resulted in a better outcome than the current state. Furthermore, are we now in a similar situation with the refugee crisis? In organizations such as the military, in which there is an imperative on maintaining the initiative through speed of operations, inertia can be even more costly. Given the general lack of practical and theoretical research in this area, this book pays special attention to the concept of inertia: why it emerges, what happens when people become inert, and what factors (at the personal, social, and organizational level) increase or decrease the likelihood that inertia will emerge.

Overall, the findings of our research surprised us. Compared to their police and emergency service counterparts, members of the Armed Forces were remarkably (although not universally) resistant to inertia: In the face of incredibly hard decisions and in incredibly hard environments, they remained able to choose and act in a timely manner. They were able to tolerate a bad (albeit least-worst) option. Throughout this book, we allude to the potential factors that could be driving this difference between soldiers and emergency service personnel (for whom we have found a great deal of decision inertia), as well as whether these factors might help train better decision-making in fields that face similar high-stakes situations.

Our research also unearthed several other interesting findings about military decision-making. First, the *types* of decisions that soldiers struggled with continued to surprise us. The decisions that were discussed, while often high-risk and involving issues of life and death for their forces and civilians, were rarely "kinetic" (meaning that they were not decisions faced in combat, although in Chapter 8 we explore in great detail one such decision), but more often they were encountered during the mission planning process. Furthermore, these decisions were often not merely "hard" because of the costs (which have almost always been linked to the difficulty a decision poses, as covered in Chapter 1). More often, they were hard because they came from "left field"—meaning that the decisions had to be made in completely unexpected situations that they had never trained for and did not expect to encounter. There are some decisions that time and time again caused soldiers difficulties, such as how to determine a threat (explored in Chapter 4), but strikingly the decisions we discussed with soldiers arose in unique situations that posed unique challenges regarding decision-making.

One of the things that also struck us throughout this research was the vividness with which soldiers remembered these types of decisions. During any operation, on any given day, on any deployment, a soldier can make hundreds (one even said millions) of decisions, yet our participants could always recall a least-worst decision in great detail. Although this provided much-needed detail to the recollection (sometimes of events that occurred more than 10 years ago), it highlights serious concerns about the enduring effects on soldiers of making these kinds of decisions. So, although they tolerated a bad outcome during goal-directed, short-term actions, are the long-term consequences potentially intolerable? In discussing this research with a trained veterans' counselor, the counselor said we were studying "shoulda, woulda, coulda" decisions—decisions that haunt the soldiers and "come back with them." From our research, we can confirm that in many cases, decision-makers continued to ruminate on the outcome of a decision years after it had originally occurred. In Chapter 9, we explore the potential relationship between least-worst decisions and post-traumatic stress disorder in the veteran population, seeking to highlight the role of decision-making—specifically, how the types of decisions made on deployment by members of the Armed Forces can have long-lasting (and detrimental) psychological consequences when they return home.

In 2015, the war in Afghanistan began its 15th year, making it the longest conflict in US history, and the collection of published literature on experiences in Iraq and Afghanistan is growing. The success of and attention to several of these books show the current appetite for understanding what being "at war" fully entails. The common thread of these books is that they portray the grim reality of war and how war can act as a "laboratory for the human condition in extremis" (Filkins, 2014). We hope that this book will help you understand the psychology behind military decision-making while keeping the salience of war at the forefront. By presenting a series of least-worst decisions faced by individuals across all branches and levels of command (from colonel to infantryman), this book undertakes a psychological analysis of decision-making that does not lose its grounding in reality. Instead, it allows the reader to imagine wearing the soldier's boots while also seeing how the psychology of such decisions played out in real life.

Dr. Neil Shortland, Prof. Laurence Alison, and Dr. Joseph Moran

ACKNOWLEDGMENTS

I always thought that writing a book would be a pain, but writing this book has been an incredibly positive experience. The main reason for this is my co-authors. Professor Laurence Alison, during the past 5 years, has been an incredible mentor and friend, and I have benefited immeasurably from our time together. I hope this is the first of many collaborations to come. Joseph Moran has been equally vital in shaping this book. His applied work with soldiers played a central role in shaping the content and presentation of this work. He also understands the rules of grammar, which, it turns out, I do not.

The majority of this book was written during my time as a Visiting Fellow with the University of Oxford's Changing Character of War Program, Pembroke College. I am incredibly grateful for my time at Oxford and all those I engaged with while I was there. Although there are many people whose expertise I benefited from during my time there, I am truly grateful to Dr. Rob Johnson, Dr. Annette Adler, and Ruth Murray for their support throughout the program.

I also thank all those at the University of Massachusetts Lowell and, specifically, those I work with at the Center for Terrorism and Security Studies. Both the faculty and students have been a huge source of support and encouragement. Special thanks go to Prof. James Forest, Dr. Andrew Harris, Prof. Luis Falcón, and Prof. Eva Buzawa.

Going wider still, I am truly grateful for the support I have received from friends and family (both in the United States and home in the United Kingdom). Some of the brainstorming for this work was also done on the golf course, so it is only fitting that I thank Ian Elliott, Benjamin Grindley, and Matthew Crayne.

Of course, my biggest thanks go to those closest to me: Josie, Mum, Dad, Jenny, Mark, Sarah, William, Thomas, and Scott.

Neil Shortland

I thank Neil for asking me to contribute to this important book. It was a pleasure and a privilege to help out and work alongside him and Joe. On the professional side, I dedicate this work to all my students—past and present—for their creativity, insights, and endurance. On the personal side, I dedicate this to my parents, John and Jean. Much of this book is about values, and any of the good ones I possess come from them.

Laurence Alison

Writing this book has been an amazing journey, and first I thank my co-authors, Neil and Laurence, not only for their immense efforts but also for their unfailing good humor in the face of my grammatical nitpicking.

I thank my former colleagues on the cognitive science team at NSRDEC: Caroline Mahoney, Caroline Davis, Marianna Eddy, Matt Cain, and Tad Brunye. They have all shaped much of my thinking about cognition in soldiers, and to them I am grateful for the many discussions that have improved my contributions to this work.

Finally, and most important, I thank my family, Tammy, Taylor, Nick, and Alex—you have been very patient while I have spent many a night and weekend working on this project.

Collectively, we all thank Dr. Joan Bossert and Lynnee Argabright at Oxford University Press, without whom this book would not exist. Thank you for seeing merit in our idea and supporting us throughout the publication process. Finally, a thanks to Colonel Tim Collins for providing a compelling starting point with a powerful, provocative, and thoughtful foreword.

Joseph Moran

1

DECISIONAL CONFLICT FROM BEST TO LEAST-WORST

Make the most of the best and the least of the worst.
—Robert Louis Stevenson

The goal of this book is to investigate both the type of decisions that are faced by soldiers when at war and the psychological process they must go through to identify, choose, and commit to a course of action. Before delving into the narratives provided by our interviewees, we believe it is necessary to reflect on what decisions are and what, critically, makes some decisions "harder" than others. The reason is that despite often emphasizing how "hard" the decisions made by members of the military must be, we often fail to actually identify what makes certain decisions hard. We seek to unpack in this chapter the difference between high-risk outcomes and hard decisions because although many hard decisions have high-risk outcomes, high-risk outcomes alone do not make a decision hard.

In the first interview I (N. Shortland) conducted with a soldier, he recalled, in hindsight, that

> I feel like I got what we call a "script save," I made the right decision for the wrong reasons and that's the way I still look back on this; I made the right decision for the wrong reason.

We explore the full context of this decision in Chapter 4 (see "The Walk-In"), but his sentiment highlights that sometimes the quality of a decision itself and the quality of the process used to make that decision can diverge. Good decision-making ≠ good outcomes, and vice versa. Thus, although we agree that there are better *outcomes*, better outcomes do not always follow better decisions (as is implied in the previous quote). This is a critical aspect of real-world decisions. In this chapter, then, we define both "hard" decisions and "quality" decisions within the military context.

WHAT IS A "HARD" DECISION?

Philosophers, mathematicians, economists, psychologists, and political scientists have all attempted to understand how people make decisions in various different domains and under various conditions (Payne, 1976; Plous, 1993; Treviño, 1986). Yates and Estin (1998; see also Yates & Patalano, 1999, p. 15) defined a "decision" as "a commitment to a course of action that is intended to produce a satisfying state of affairs." This highlights two facets of decisions: a resulting action and the quality of the action. The quality of a decision outcome (after the fact at least) may be relatively easy to compute (lives lost, damage incurred, etc.), but understanding why decisions are hard is itself more difficult to quantify. Yates, Veinott, and Patalano (2003) asked 93 undergraduate psychology students to "imagine hard and easy decisions that you have made within the last year" and write down three hard and three easy decisions. They were asked to describe the circumstances around the decision, explain why it was hard (or easy), and then outline how they solved a given problem. Yates et al.'s research resulted in an outline of 212 hard decisions and 200 easy decisions. Unsurprisingly, undergraduates often thought that academic concerns (e.g., "What college should I attend?"), social issues (e.g., "Who should I date/not date?"), and financial matters (e.g., "Should I buy a car and, if so, which one?") led to "hard" decisions. Yates et al. used the responses to cluster seven "supercategories" that defined why decisions were difficult:

Hardness factor 1. *Outcomes—Serious:* Decisions in which the outcome could result in a "serious loss of some kind." Participants most often cited losses that were long term and had potentially irreversible effects. These often entailed hurting another person, or ones that required violating personal (e.g. moral) principles. Outcomes that involved significant risks were also within this supercategory.

Hardness factor 2. *Options:* Decisions with either too few or too many options, as well as those that involved too many factors, were viewed as hard.

Hardness factor 3. *Process—Onerous:* Decisions were hard if the amount of effort required to make the decisions was high and/or if any of the following were present: uncertainty, time pressure, emotional issues, or a feeling that they (the deciders) lacked expertise.

Hardness factor 4. *Possibilities:* Decisions were viewed as hard if it was difficult to imagine what the possible outcomes could be. This was especially relevant for domains in which the decision-makers had little experience of the type of decision they were making (e.g., buying a first house/car or choosing a college).

Hardness factor 5. *Clarity:* Decisions were viewed as hard if it was unclear *which* possibility was superior in relation to the others it was being compared against.

Hardness factor 6. *Value:* Decisions were perceived as hard if the decision-makers were unsure how they would *feel* about a given outcome. For example, they may have never experienced an outcome (e.g., being a doctor or lawyer) and therefore they were unsure whether (and how much) this would be a pleasurable outcome and worth the costs (several more years of college and incurring increased debt).

Hardness factor 7. *Advisors:* Participants said that decisions became hard when they had opposing advice or when advice contradicted their preference.

Conversely, "easy" decisions were (logically) defined as those with moderate outcomes, a clear optimal option, little effort, predictable outcomes (and predictable personal responses to these outcomes), and the ability to defer to (and follow) an advisor.

Let us examine these factors with a scenario we designed to elicit maximum decision-making difficulty,

---

THE OFFICE PARTY SCENARIO

Imagine that you are at an office party with your work colleagues or people you engage with in a professional context (e.g., an academic society social). Present at this event are individuals you know and have worked with for many years but who are employed by other organizations (or agencies). You have had a few drinks (but certainly are not drunk), and you have just finished the main meal and are waiting to order dessert. Now imagine someone you have known for many years and would consider yourself pretty close to (but is not necessarily part of your "team" or institution). If you've now got this person in mind, imagine that they were sat next to you at this event. This person has just returned from the restrooms (or "toilets" to UK readers). He or she tells you something about one of the other staff members at the meal who is seated on the opposite side of the table (again, select this person to be someone you know well but in this case is also *part of your team*; we refer to this person as "person X"). The person next to you says the following: "I've just seen person X in the toilets. S/he was leant over a sink and then when s/he saw me, s/he put what looked like a see-through small bag with white powder back in their pocket. S/he then said, 'for God's sake please don't say anything' and then quickly left."

Let us relate this example to the seven "hardness" factors identified by Yates and colleagues (2003). First and foremost, the outcomes are very serious. From the little that is known, there is the potential that an individual working in your organization is using cocaine. The outcome for the subject of the decision ("person X") is significant either way: If you intervene, person X may be at risk of losing his or her job; if you do not intervene, perhaps person X does not get the help that he or she needs (resulting in increased use, which could eventually lead to all sorts of negative consequences). The consequences for you, the decision-maker, are also significant in that reporting this behavior leaves you at risk from backlash from friends of person X, but not reporting it (and being discovered for not doing so) leaves you in jeopardy of organizational backlash for (potentially) seeing an illegal act and doing nothing. In relation to hardness factor 2 (Options), there are many options to compare along many different dimensions. First, you must make the binary decision to do something or to do nothing. If you decide to do something, you must then decide what to do (tell a friend, tell your boss, tell person X's friends, or tell person X's boss), and once that decision has been made, there are still several more decisions to be made about what you tell person X (do you tell him or her that you heard it could have been cocaine or simply that you heard he or she was using drugs?). In terms of hardness factor 3 (Process—Onerous), it is certainly onerous in that the decision is highly emotionally charged, there is time pressure, and there is much uncertainty. In addition, it is very difficult to foresee what the outcomes could be (hardness factor 4), and there is little clarity regarding which of the many possible options is superior to others (is it better to get person X the help he or she needs, ensure person X does not get fired, or ensure you do not get fired?) and how you may feel about the outcomes that may stem from your actions (factors 5 and 6). Finally, you currently have no advice, and in fact the process of seeking advice is a decision in and of itself (factor 7). This very simple scenario therefore encompasses all of the hardness factors proposed by Yates and colleagues.

In their original study, Yates and colleagues (2003) coded for the presence of hardness factors within the 412 decisions reported. They found that the most common reason a decision was viewed as "hard" was because there were serious outcomes to the decision. Thus, even as research aims to understand why the *process* of decision-making can be difficult (i.e., choosing a course of action and implementing it), people think that the *outcome* is more critical. This finding has two implications. First, it belittles the challenges that are posed by the choice itself, instead shining extra light on something that already has the spotlight. Second, it means that people believe that all decisions with serious consequences are ipso facto hard. Although this makes intuitive sense, it does not reflect that in many high-risk and high-consequence situations, individuals across a number of fields can make decisions with relative ease. Furthermore, Yates et al.'s research found that hard decisions demanded more time to make, obviously implying that decisions with high-risk outcomes take more time. In the real world, then, this puts a premium on being able to make these hard decisions in situations that afford little time. For example, in 1824, the surgeon James Bonnar wrote the following:

> A hesitating practitioner, who takes a few hours only to make up his mind respecting the course he is to pursue in the treatment of a doubtful disease, may often thus doom his patient to an irretrievable state. I hold it as a maxim, that it is more culpable for a physician to lose his patient by neglect or indecision when it might be in his power to save him, than it is one who hurries on the fatal termination by the use of desperate measures in a desperate disease. (p. 33)

Thus, outcomes cannot be the only factor that marks decisions as "hard," and instead decisions should be considered "hard" not solely because of outcomes but also because of many factors that influence the decision process.

DECISIONS IN "HARD" ENVIRONMENTS

In our own work with emergency services, two factors continually separate hard choices from easy ones: task ambiguity (not being clear on exactly what is the task at hand) and outcome uncertainty (not being clear on what will be the precise consequences of one's decision). This can be mapped out in two orthogonal dimensions (Table 1.1).

Consider Table 1.1 as part of an imagined "bomb" paradigm in which the options available to you are cutting a red or blue wire to disable a bomb. When there is both task ambiguity and outcome uncertainty (condition D), there tends to be a fierce internal struggle to commit to a course of action. The experience of decisional conflict

Table 1.1 Task Ambiguity and Outcome Uncertainty with Reference to a Bomb Paradigm

	Clear Task	*Ambiguous Task*
Certain outcome	**A.** Cut red wire disables bomb.	**B.** Red *or* blue wire will disable bomb (not sure which).
Uncertain outcome	**C.** Cut red wire may disable or may detonate bomb.	**D.** Red or blue wire may (or may not) disable or detonate bomb.

("If I cut the wire, that might be good or bad and if I do not cut the wire that might also be bad. Also, which wire do I cut?") can be exacerbated if the decision-maker is told that the bomb is somewhere in a preschool (as opposed to, for example, an empty car park) because, of course, the consequences are much higher—we do not completely discount the influence of serious outcomes. Unfortunately, just such ambiguous, uncertain, and high-stakes consequence factors represent the reality of many critical and major incidents. In these cases, it can be very difficult to establish situation awareness in order to reduce task ambiguity, and it can be harder still to then decide which of several unknown outcomes is the least-worst.

One of the key reasons we have been able to describe extremely difficult decision environments is because of the opportunity we have had to speak to practitioners and hear from them what they found challenging and why. All too often (in our view), researchers start with what they think are the challenges without the common-sense approach of getting that answer from the horse's mouth. Getting the horse to speak can of course be extremely difficult: The multiagency, multijurisdictional, and often political environment can frustrate access to honest and detailed exposition from those individuals who made decisions in high-stakes situations such as the emergency services or law enforcement. To solve this issue, Jonathan Crego (who has worked alongside Prof. Alison for many years), at the behest of the Association of Chief Police Officers, developed an anonymous reporting system known as 10,000 Volts (10kV—a moniker developed after one senior officer described his first critical incident in the role of commander as feeling like being hit by 10,000 volts). The system was based on work in the aviation industry, in which air traffic controllers wanted an outlet to express near misses as a product of systemic failings rather than personal errors. It was believed that to really access what was going on in these cases, air traffic controllers needed free reign to say what they wanted without fear of personal repercussions.

Commonly in 10kV, the key players in a critical incident sit in a horseshoe configuration around tables with laptops or tablets, and each subsequent debriefing prompt (What issues could have been better handled?) emerges on the screen, enabling participants to write their own views. All views are then disseminated to all terminals, but no one knows who wrote which comment (hence anonymous and not source attributable). An advantage of 10kV is that both science and intuition tell us that people impression-manage in face-to-face discussions: Some do not bother to contribute because the loud voices dominate, some worry about upsetting the boss or revealing personal issues, and others do not listen because they dislike the speaker. 10kV's anonymity eliminates such concerns. Ideas are judged on their own merits, not on the merits of the speaker or of his or her rank. Controversial or sensitive topics are more easily discussed (e.g., admitting personal trauma or stress), and mistakes can be admitted blamelessly. Research supports these claims. Simply, people are more honest when reporting is anonymous. Experience of 10kV shows that people give very candid, frank accounts, and many state that they find it "cathartic" (Alison & Crego, 2008, p. 37).

We have conducted and analyzed more than 450 10Kv sessions, including with the individuals involved in responding to the Buncefield fire (Europe's largest peacetime fire, which erupted after 11 explosions at an oil storage facility), the bombings in Sharm El Sheik, the response to the Boxing Day tsunami, the preparations for the London 2012 Olympic Games, as well as the response and the aftermath of the 7/7 London bombings. In particular, during these debriefs, delegates consistently referred to the following issues that emerge and made their job as decision-makers difficult time and time again:

- Volume of information (too much, too little, or contradictory)
- Uncertainty and stress (misinformation and high-stakes, dynamic, and fluid scenarios)
- Inability to diagnose the problem—sense-making (interpreting cues correctly and quickly and effectively acting on them; recognizing redundancy and errors and correcting them expeditiously)
- Multiagency communication (competing agendas, information sharing, territories, organizational stovepipes)
- Emotion (career fear, regret, pride, loss, and safe self-cognitions)
- Confidence (in team and self) and legacy (for victims, community, organization, and self)

Revisiting Yates et al.'s (2003) list of "hardness" factors, it is clear that many of these are reflected when we examine decisions made in the real world. These decisions are "onerous" in that there is high uncertainty (a lack of ability to "make sense" of what is going on, what has happened, and, crucially, what will happen), there is emotion (surrounding the event and the consequences for you, the public, and your colleagues), and there is often a degree of time pressure. Given this uncertainty, it is difficult to determine the "possibilities" and imagining or predicting what will happen. This makes determining any "superior" course of action incredibly difficult. In addition, you have to juggle the input of "advisors" (even superiors) whose priorities and values may not match yours.

By way of illustration, let's consider the response to the 7/7 bombings in London in 2005 (the response to which was referred to as Operation Theseus). The 7/7 bombings were a series of coordinated suicide attacks on the London Underground in which three tube stations were attacked (Aldgate, Edgware Road, and Russell Square), as well as a bus at Tavistock Square. Fifty-six people were killed, including four bombers, with a further 700 treated for injuries. Operation Theseus was one of the first debriefs Prof. Alison had been involved in, and it certainly was a wake-up call in reaffirming his oft-espoused views about problems of narrowly confining oneself to laboratory-based, tightly controlled research. From this debrief, it was clear that 7/7 pressed many of the "difficulty buttons" we identified previously—problems of misinformation, uncertainty, and stress and difficulties in both situation awareness and correct identification of what one was dealing with. Consider some of the comments from those who were there:

> Sitting at [tube network control] 8.52 am, you are virtually blind and you are confused for a while as these multiple reports come in. [We cannot] pretend that we have instantaneous appreciation of what is happening. We do not, and the reports that come in conflict with one another.

Because of a loss of power and the loud bang, a power surge on the line was initially assumed. Then, because of smoke issuing from tunnels, a train crash was assumed. Although emergency services responded within minutes at 9:15 a.m., ambulances were initially sent to seven different sites because command and control, although it now knew *what* had happened, did not know *where* the explosions had occurred.

Two comments from survivors on the trains illustrate the frightening lack of situation awareness:

> We waited for help, that was the main concern, if there was smoke, there must be a fire on its way, burning down the tunnel towards us. If people had known there was no fire (through someone making contact with us) the situation could have been a lot calmer. I think the most important thing that needs to be recognized is us not having contact with anyone.
>
> To our horror we then heard a train, thinking it was coming towards us, people were screaming there was a train coming towards us and that no-one knew we were down there. That was the scariest part of it (apart from thinking I was going to burn alive)—not knowing.

Thus, 7/7 was a clear example of a chaotic event in which situation awareness was frustrated by volume of information (too little and contradictory in the earliest stages and then a proliferation as people emerged from the underground); uncertainty and stress (misinformation about the nature of the explosions); high stakes (insofar as people were dead and dying); dynamic and fluid scenarios and an inability to diagnose the problem (i.e., they could not make sense of the problem—were there four or six bombs? And if they could not diagnose the problem, how could they effectively react to it?); multiagency communication (competing agendas between crime scene forensics and the need to get in the tunnels and save lives; information sharing concerning different systems, communication systems, and protocols; etc.); emotion; "career fear" (worrying about the outcome that a decision will have on one's job security); anticipatory regret insofar as any decision could easily be the wrong one; confidence (in the team and self); and legacy (for victims, community, organization, and self). As such, 7/7 included both task ambiguity and outcome uncertainty—situation D from Table 1.1—the worst possible combination of factors. Later in this book, we bring to life difficult decisions by letting military decision-makers explain in their own words how real-world decision points conspire to stack these hardness factors atop one another.

LEAST-WORST DECISIONS

Based on the previously discussed findings, and others emerging from our research on real-life critical incidents, we have become increasingly interested in what we call "least-worst decisions." What defines least-worst decisions is that *all* courses of action are adverse: Every course of action is high risk, and every course of action leaves us facing negative consequences. Such decisions pose a significant problem to current perspectives on decision-making, which hold that there is an *ideal, best*, or *workable* solution to a problem.

Those who study organizational decision-making have long preached optimization: selecting courses of action that provide the greatest payoff (e.g., Svenson, 1979). Such models inherently assume both complete information and human rationality, proposing that humans make decisions knowing what all the payoffs are and based on expected benefits alone (this concept is discussed in detail in Chapter 2). However, as Simon (1976) so succinctly observed, humans do not have "the wits to maximize" (p. xxviii). As Simon argues, determining all the (potential) favorable and unfavorable consequences of all feasible actions would overstretch the limited cognitive capabilities of humans. To solve this theoretical conundrum, Simon (1955, 1956, 1958) proposed that decision-makers "satisfice." This approach bypasses our cognitive limitations by substituting one task for another; working to a threshold of acceptability instead of searching for and selecting the optimal choice. Once the decision-maker finds an option that exceeds this threshold, it is chosen. Such strategies are beneficial when the mental effort savings outweigh the foregone payoff costs (Simon, 1955, 1956). Satisficing is supported by the popular psychological view that we are "cognitive misers" who dole out cognitive effort as minimally as possible (Fiske & Taylor, 1991; Gonzalez, Dana, Koshino, & Just, 2005). This kind of perspective is seen in the work of Kahneman (2011), who argues persuasively that much of our behavior is governed by what he calls "System 1," the nonconscious, associative, cognitive neural network that constantly analyzes the environment and incoming information and serves up plans of action fully formed to consciousness. In contrast, we only engage "System 2," what we might experience as "thinking on a problem," when absolutely necessary—that is, when System 1 does not give us a satisfactory answer.

Satisficing is a central facet of current perspectives of decision-making, but least-worst situations often present with no "workable" options, or they present options that are equally "unworkable" (none of which would be viewed as "ideal" and all of which have an unpalatable outcome). Furthermore, whether a course of action meets a "threshold of acceptance" is often, in high-risk adaptive environments, only known *after* the decision is made and the chips have fallen. It is in situations such as these that decision-makers, typically reliant on satisficing, struggle to select. In Chapter 2, we discuss in detail the recognition-primed model of decision-making (RPD; in which previous experience helps identify workable courses of action). However, such perspectives are virtually silent on what to do when none of the options are workable. RPD researchers call facing equally averse options the "zone of indifference" in which "the closer together the advantages and disadvantages of competing options, the harder it will be to make a decision, but the less it will matter" (Klein, 1998, p. 107). As such, decision-makers should "stop right there, make an arbitrary choice, and move on" (Klein, 2011, p. 87). But when in the zone of indifference in high-stakes decision-making environments, decision-makers rarely make an "arbitrary" choice. Instead, decision-making in these circumstances is usually derailed (Alison et al., 2015; van den Heuvel, Alison, & Crego, 2012). Moreover, the term "arbitrary" sits uncomfortably with regard to, for example, shoot/no-shoot decisions, the decision to deploy troops to Syria (or not), or the decision to accept X or Y number of refugees per European country. These difficult-to-calculate options, in which it is difficult to discriminate the least-worst long-term outcome, cannot be considered arbitrary—they are fundamental, life-shaping events for victims, communities, countries, governments, and soldiers—the risk is too great to allow a coin flip to guide us out of the zone of indifference. It is here where the RPD approach falters because it is painfully obvious that the outcome of the decision could not matter more to the parties involved.

Let us consider, for example, the strategic-level decision-making of President Obama surrounding intervention in Syria. Although many have been quick to lament the indecision of the Obama administration, even a quick look at the options available reveals that none offered much promise. As (then Secretary of Defense) Hillary Clinton (2014) stated,

> Do nothing, and a humanitarian disaster envelops the region. Intervene militarily, and risk opening Pandora's box and wading into another quagmire, like Iraq. Send aid to the rebels, and watch it end up in the hands of extremists. Continue with diplomacy, and run head-first into a Russian veto. None of these approaches offered much hope of success. (p. 461)

Although the case of Syria represents a strategic-level least-worst decision, these types of "damned if you do; damned if you don't" decisions also emerge at the tactical level, especially in a wartime situation. Consider, for example, the following situation encountered by a Marine officer (recounted in Friedman 2007):

> Suddenly the guy reached down and picked the object up. "Hey what the fuck! Drop that motherfucker!" I screamed, raising my weapon. I started moving backward. Mohamed started shouting at the guy and the guy started talking back to him, all the while smiling and shaking the yellow object. . . . In my mind, I quickly started to work out the combat calculus that would tell me how to deal with this. . . . I didn't want to have to shoot this guy over a misunderstanding, but dying for this guy's stupidity was not an option either. This was the same thing that happened to Charlie Company and the Iraqi who died right in front of me. Given the choice I would shoot first and ask questions later. This was not how I was going to die. (pp. 162–163)

Although this presents a simple tactical shoot/don't shoot scenario (i.e., there are only two options), neither option is workable: Either (1) shoot an unarmed civilian or (2) put his and his fellow soldiers' lives at risk by not taking offensive action. This type of situation (cultural misunderstandings leading to a least-worst decision) emerged often in Afghanistan and Iraq given the significant cultural differences and language barrier between troops and the civilian population. What makes these strategic (intervene/don't intervene) and tactical (shoot/don't shoot) decisions similar is that there are only high-risk courses of action, all of which could have negative consequences for the decision-maker, his or her organization, and civilians. In such instances, the decision-maker *must* select the least-worst course of action; there is no "best." Furthermore, in most highly time-sensitive military situations, not opting is not an option. This non-availability of suitable alternatives puts the decision-maker in a state of decisional conflict (Zeleny, 1976).

DECISION ERROR: OUTCOMES VERSUS PROCESSES

In the definition of decision-making we proposed at the start of this chapter, we highlighted that decisions are intended to produce a satisfying state of affairs. However, often, "better" decisions are only known in hindsight, and we must not discount the role of external (uncontrollable) factors and luck. As such, we advocate that a more profitable path is to focus on better decision *making*—as a process rather than an outcome. This is especially relevant, we argue, in military settings. In Chapter 7, for example, we discuss a decision made by Lt. Col. Davids,[1] who was required to negotiate the rescue of a military asset (he refers to this as a "scraper") that had been targeted by an insurgent attack. In reflecting on his decision-making, he highlighted the importance of ever knowing if outcomes will be good:

> You never know if it is the right decision or not. Someone could have parked a car bomb right there on the side of the street next to it, you could have had multiple casualties, stuff like that was running through my mind. Stuff like that was running through my mind, you know, what is the worst thing that can happen? You see it might be, plugging the cobalt beneath the street full of explosives and just waiting for us, and just baiting us with that. And all that stuff was certainly in the back of my mind. It was those kinds of risks. And then someone would say, man that was totally the wrong call, putting that many people in that kind of an area, you just created this huge target. And you know, was this scraper worth it? So that was always there. But I always felt like, and I probably felt on most of my missions that I always did everything I could do to mitigate risk.

Lab-based decision-making research has served as the workhorse, but such studies do not capture the shifting calculus of the real world. The decisions posed to participants in a lab, for example, often have a "wrong" or "worse" answer, from which we can infer that someone engaged in an "incorrect" decision-making process. For other types of decisions, such as multiattribute decisions, we can infer a decision is worse if it does not match the needs and goals that the decision-maker had at that time. Consider the common multiattribute decision of buying a car. Here, we juggle with several different criteria: numbers of seats, price, fuel economy, and insurance costs. The best decision here is thus the one that, overall, has the highest evaluation across all variables. In such circumstances, an error (or worse decision) is

---

[1] Throughout this book, the names of all interviewees (and any others mentioned) have been changed to protect the anonymity of the interviewees. In many cases, specific times and locations have also been changed or removed.

defined as choosing an option that had a "lower score" than an alternative that was available to the decision-maker at that time. For example, one of us (N. Shortland) recently purchased a Ford Mustang, which does not constitute an "error." However, buying this car if his wife was pregnant with twins might constitute a worse decision (in her mind definitely, even if not in his). Buying a two-seater sports car when one has a family of four may also constitute a worse decision. Such approaches to error, however, do not work in complex environments with their insistence on uncertainty, ambiguity, and paucity of information.

As psychologists continued to study decision-making, it became increasingly apparent that the way people made decisions "in the lab" did not capture how they behaved in the real world. Hence, naturalistic decision-making (NDM) emerged as a theoretical and practical approach "to understand how people make decisions in real-world contexts that are meaningful and familiar to them" (Lipshitz, Klein, Orasanu, & Salas, 2001, p. 332). The assumptions of NDM and the main theories in this field are presented in detail in Chapter 2. However, how NDM researchers treat "errors" is worthy of discussion here. Contrary to lab-based work, NDM rejects the notion of "right" and "wrong" decisions. Instead, NDM researchers seek to understand the cognitive processes associated with decision-making "in the wild" (Gore, Banks, Millward, & Kyraikidou, 2006). Using the naturalistic perspective, Orasanu, Martin, and Davidson (1998) sought to explain the role of decision processes in negative outcomes. As Lipshitz et al. (2001) argued, "We need to consider the consequences of errors, not just the reasoning processes" (p. 340). This approach consciously explores the relationship between negative outcomes and negative decision-making processes rather than just assuming negative outcomes imply bad decision-making processes. As such, outcomes per se cannot be used as an indicator of decision-making quality. This is what Baruch Fischhoff and others have referred to as "hindsight bias"—the propensity to define errors by consequences (Fischhoff, 1975; Hawkins & Hastie, 1990). With time ticking away in the Super Bowl and your favorite team losing by less than a touchdown, do you believe it is a good idea to throw into the end zone from the 1-yard line when the team has the best running back in the league? In hindsight, most observers decided that Pete Carroll and the Seattle Seahawks made a poor decision that resulted in an interception and losing the 2015 Super Bowl to the New England Patriots. But just how many people made that call while the ball was still in the quarterback's hands prior to the pass? In war, adverse outcomes can be a case of bad luck as much as bad decision-making (Orasanu et al., 1998). Hence, throughout this book, as we delve into the decisions made by real soldiers, "errors" are viewed as deviations from ideal *processes* of decision-making, and not the outcomes of the decisions themselves.

CONCLUSION

Although we may not have time to be optimal in the real world, how do we know when someone has made a bad decision? In our view, part of what makes a decision "good" is that it often means committing to a least-worst or most tolerable outcome *given the circumstances*. Moreover, as we argue in Chapter 3, "bad" decisions are sometimes about a failure to act quickly enough or at all. "Overthinking" can lead to the problem running away from the decision-maker. Paradoxically, then, thinking can, under some circumstances, be a bad thing. Our view is that expertise and experience matter; ground truth matters; improvisation and creativity matter; and timing and propulsion in action and moving forward matter.

2

MILITARY DECISION-MAKING DOCTRINE, RATIONALITY, AND FIELD-BASED APPROACHES

In a situation where the consequences of wrong decisions are so awesome, where a single bit of irrationality can set a whole train of traumatic events in motion, I do not think that we can be satisfied with the assurance that "most people behave rationally most of the time."
—C. E. Osgood

The most extreme scenario was north of Konduz, Afghanistan. There was a village up there called Barouc and we knew the Taliban were amassing in this town and estimates were in the low thousands. . . . So we sent in one of our Special Operations Forces [SOF] teams to get into position to get a better sense of who was in there, maybe call in some airstrikes on some of the perimeter elements that were isolated from this town to minimize the collateral damage and also just to probe to see what was going on in there to see if we could draw them out. So we had essentially four or five guys. A couple of Americans and a few Northern Alliance guys driving up in close proximity a few kilometers away from what we thought was a couple thousand Taliban. Not very good odds, but they had a decent escape route and thought it was relatively clear. So they go in, isolate a couple of areas that are confirmed Taliban, call in a couple of airstrikes; all hell breaks loose. The enemy pretty much determined where the airstrikes were being called in from and decided to attack. This team had been working a couple of fighters [planes] as they were probing these targets and essentially the next radio call is

"Oh geez, here they come,"

"How many?"

"There's a thousand to fifteen hundred, all of them coming at me"

"How far away are they?"

"Three kilometers"

"What's their speed?"

They are passing this information back and forth.

"Okay SOLAR [the Air Support Operation Centers' call sign] what have you got for me? I need something right now?"

"We got another formation of fighters"

"That's not going to cut it, I need something to take out essentially a grid."

The formation of this valley as they were coming up to the post nicely funneled all these guys coming at him. He saw this as hitting a home run, and said "I'm going to stay here and set this trap, draw them up to me." We told him we had B-52s available but it was going to take 20–30 minutes for them to get there. He said "*Great*... what have you got in the meantime?" He got a couple more fighters and we did some harassment bombing. He built a bomber box, set up like a grid pattern to match this area in this valley, he did a very nice job of forming it. We put him in contact with the B-52s and they did their coordination, but as they were getting closer, from the tempo and the excitement in his voice, you could tell things were getting dicey. He said "Okay I can only stay here for 10 more minutes. Oh, I'm taking some shots from some guys coming in from the side. I can only stay 5 more minutes to control this air." Then the last transmission was essentially "I don't know if I can get out of here, I think my escape route might have been cut," so his last request was "If you do not hear from me, bomb the bomber box that I gave you." The implication was "They got me, and you might as well bomb it anyway, because it is the end of the game for me." So at this point we get the commander, explain to him the situation and tell him that this guy had requested to execute this mission even without his final control because the mere fact that we had not talked to him tells us he has probably been captured or killed. This is a very tough decision for the commander, and it is his alone to make. So a pretty tough three or four minutes. Seemed like an eternity, as the airplane is checking in with no contact from the ground, and the last thing you hear is kind of like ricocheting and bad stuff going on. So we are all sitting there, your stomach is getting tied up in knots, the whole tent is silent and all eyes on three or four folks staring at the radio with the commander making this decision. Finally, he said "Alright, do it." We relayed to the B-52s, "Mission's a go. Clearance given on our order, no final contact required." So the mission goes. We still don't hear anything from the SOF team. Every minute that goes by we think we probably just killed our own guys. Probably ten minutes later we get a radio call "SOLAR, SOLAR, team XX, thank you, thank you. We are all fine." And that is as close as I ever want to see it again.

This is just one example of the types of decisions that had to be made by someone during the war in Afghanistan. In this case alone, there are many of the factors that we identified in Chapter 1—principally that there was no "right" or "wrong" answer, and the outcome, which was, fortunately in this case, a good outcome, could have very easily been disastrous independent of the decision-making process that the soldier went through. With this case as a backdrop, in this chapter, we explore the doctrinal and theoretical perspectives currently employed to (1) help the military make decisions during operations and (2) help us understand the process through which such decisions are made and the cognitive factors that underpin these types of decisions. We discuss three broad approaches to understanding how military decisions are made: (1) the military decision-making process (MDMP), based largely on historical experience and, arguably, best described as the nearest thing to a doctrine-based approach; (2) rational–cognitive approaches based on observations in the lab; and (3) so-called "naturalistic decision-making" approaches based on observations of decision-making in the field. We show how the strengths and weaknesses of each approach reveal considerable gaps in our knowledge about the specific nature of the impossible decisions that soldiers face.

In Chapter 3, we summarize our own developments built upon each of these approaches. Our SAFE-T model (*s*ituation *a*ssessment, *f*ormulate plan, *e*xecute plan, *t*eam learning) has the benefits of phased-based models (aligning with what naturalistic decision-making [NDM] has identified as a common feature of real-world decision-making) but is far more streamlined than the MDMP. In addition, it recognizes the very real "error spots" within and across these phases—spots that have much in common with rational–cognitive approaches to decision-making that have identified how and when error occurs. As such, SAFE-T is a hybrid model between currently recognized doctrine, naturalistic approaches, and rational–cognitive models that warn against decision error. Moving forward throughout the book, our interviews help illustrate the nuances of decision-making within each of the phases identified in the SAFE-T model. Before we do that, let's explore the three strands of thinking that we have drawn on to inform its genesis.

THE MILITARY DECISION-MAKING PROCESS

The MDMP is the rational–methodological tool used by military personnel to solve tactical problems and make military plans. As such, it represents the Army's formal methodology for making tactical decisions (Burwell, 2001). When followed correctly, it should "lead to the best (or at least a better) decision given the degree of uncertainty and complexity of the situation" (Allen & Gerras, 2009, p. 79). The MDMP stems from the original Army process of "estimating the situation" (Michel, 1990), the first documented instance of which Maj. Von Steuben produced for General Washington regarding how to attack British forces at Stony Point during the Revolutionary War (Hittle, 1975). The process of estimating the situation became formalized in a 1910 Field Service Regulation prescribing that

> To frame a suitable field order the commander must make an estimate of the situation, culminating in a decision upon a definite plan of action. He must then actually draft or word the orders which will carry his decision into effect. An estimate of the situation involves a careful consideration from the commander's viewpoint of all the circumstances affecting the particular problem. In making this estimate he considers his mission as set forth in the orders or instructions under which he is acting, or as deduced by him from his knowledge of the situation, all available information of the enemy (strength, position, movements, probable intents, etc.), conditions affecting his own command (strength, position, supporting troops, etc.) and the terrain insofar as it affects the particular military situation. (p. 59)

The MDMP today reflects generations of experienced officers' adaptation since the original doctrinal conceptualization (Michel, 1990). The current MDMP, as detailed in ATTP 5-0.1 (*Commander and Staff Officer Guide*, Department of the Army, 2011), is an

> iterative planning methodology that integrates the activities of the commander, staff, subordinate headquarters, and other partners to understand the situation and mission; develop and compare courses of action; decide on a course of action that best accomplishes the mission; and produce an operation plan or order for execution.

This version of the MDMP (as with previous versions) contains seven steps, each of which begins with certain inputs and results in certain outputs. As such, the MDMP is a linear process in which each step is sequential, building on the previous steps.

The first step in the MDMP is the receipt of the mission, after which the staff prepares for mission analysis and conducts a quick initial assessment that determines (1) how much time they have; (2) how much time they need to plan, prepare, and execute the mission; (3) the intelligence they have and/or need;

(4) the manpower available to them and the manpower required; and (5) staff experience, cohesiveness, and level of rest and stress. The second step is then mission analysis, which allows the commanders to "gather, analyze, and synthesize information to orient themselves on the current conditions of the operational environment" (Department of the Army, 2011, ATTP 5-0, 4-25). Course of action analysis follows next. A course of action is a "broad potential solution to an identified problem" (Department of the Army, 2011, ATTP 5-0, 4-79). Each course of action must be suitable to accomplish the mission and comply with the commander's guidance, feasible in that the unit must have the time and resources to accomplish the mission, acceptable in that the course of action must justify the costs (including potential civilian harm) and resources available, distinguishable in that each course of action significantly differs from each other, and complete in that the course of action must represent a complete mission plan. Course of action development is an eight-stage process. The courses of action are then analyzed via war gaming (the examination of a conflict plan in an artificial environment and identifying the likely reactions, impacts, responses, costs, and benefits of a plan; Pech & Slade, 2004), which allows the commander and his or her staff to determine "how to maximize combat power against the enemy while protecting the friendly forces and minimizing collateral damage" (United States Army, 1997, Field Manual [FM] 101-5, 5-16).

Acknowledging that military operations are often conducted in time-sensitive situations, and that operations may "outrun" an initial plan, both the Field Manual 101-5 (United States Army, 1997) and ATTP 5-0.1 (Department of the Army, 2011) offer a series of guidelines for conducting a restricted MDMP. In military planning, time is both nonrenewable and the most critical resource, and the MDMP is abbreviated whenever there is insufficient time for thorough application of the full process. However, within time-restricted planning, the process *itself* is not altered. As the Field Manual states, "There is still only one process, however, and omitting steps of the MDMP is not the solution" (United States Army, 1997, FM 101-5, 5-27). There are four primary time-saving techniques. The first is to increase the involvement of the commanding officer, allowing him or her to make hands-on decisions during the process rather than being briefed (and then providing feedback) on the results. The second is for the commander to be more direct in the guidance provided, limiting the number of options that can be generated. The third is to cap the number of courses of action that can be developed (minimizing the amount of war-gaming). Finally, parallel planning can be used. This abbreviated version of the MDMP maximizes time available while facilitating adaptability in the face of a rapidly changing situation (United States Army, 1997, FM 101-5, 5-28).

PRACTICAL AND THEORETICAL ISSUES WITH THE MDMP

The military has adhered to a model of decision-making (both doctrinally and in training) that is both rational and linear (Allen, Coates, & Woods, 2012) but that is based on the experience and expertise of many generations of experienced commanders. As such, it is in line with approaches that celebrate and recognize that decision-making is phase-based, experience-led, and uses previous patterns to deal with new situations. Furthermore, the MDMP recognizes the value of scenario-based learning and the use of alternatives to test assumptions. The MDMP can be trained and encourages detailed, thorough planning and analysis of possible courses of action and the intended and unintended consequences. The MDMP has numerous clear strengths.

However, there are several pragmatic concerns with the MDMP. First, it is very time-intensive (Matthews, 2013, p. 55), it being common for commanders to invest the majority of time developing courses of action (Antal, 1998). As such, once a course of action has been approved, there remains little time to develop a full plan that includes contingencies and follow-on actions (Shoffner, 2000). The process "typically demands more time to perform ... than is afforded by the situation" (Fallesen, 1993, p. 12). Second, the MDMP rests on the assumption that the commander has good information and fully understands the battlespace (Matthews, 2013, p. 55). Third, contrary to multi-attribute decisions (e.g., buying a car), the target in military decisions is both adaptive and potentially unpredictable. Dealing with an unpredictable, adaptive opposition increases "noise," thus increasing the likelihood that a decision outcome will be poorly matched to the operating environment. This is especially pertinent when the outcome of the MDMP is a single course of action, or plan, that is optimized to work against the most likely course of action that the enemy will take (United States Army, 1997, FM 101-5; see also Shoffner, 2000).

In brief, the MDMP is labor-intensive, extremely difficult to enact in time-critical situations, and has little empirical support—either in terms of its actually being observed in real settings or in terms of whether it would be useful. That is, it neither descriptively catches what does happen nor prescriptively catches what *should* happen.

TRADITIONAL DECISION-MAKING

Practitioners of war, not cognitive scientists, developed the MDMP (Matthews, 2013), and so we might reasonably ask whether the MDMP reflects the psychology of real decision-making in the field. The problem, then, is that the MDMP perhaps reflects the early assumptions about human decision-making—that we are rational and calculated—but it has not been updated alongside our understanding of decision-making and decision-makers—that our decision-making is rough, based on shortcuts, and hence prone to errors and biases.

The MDMP is closely aligned with the "rational comprehensive" model of decision-making because it focuses on identifying alternatives and comparing them to a prescribed set of criteria (this is also referred to as "economic rationality"; Simmons, 1997). This approach is rooted in views that decision-making is algorithmic and idealized, sharing commonalities with what psychologists refer to as "traditional" or "classic" models of decision-making. "Traditional" theories of decision-making have been under development for more than 300 years and have their roots in economics, philosophy, and mathematics (Doyle & Thomason, 1999). The traditional approach is analytical (contrasted with *intuitive*), in that it assumes decisions are made through a process of logical (i.e., unbiased) probabilistic analysis. This approach has been characterized by several main concepts (Funder, 1987; Gigerenzer & Todd, 1999):

1. There is a choice between multiple available alternatives.
2. Decisions are the result of a deliberate analytical process that involves a comprehensive search for information culminating in optimal performance.
3. Models can be developed, and tested quantitatively, that will predict decision-making.

By the late 1960s, however, psychologists had amassed considerable evidence documenting numerous decision-making anomalies that derived from faulty (and definitely nonrational) reasoning (Goldstein & Hoggarth, 1997). From this, it was clear that people are not skilled "probability estimators" who derive their judgments systematically from mathematical calculations. It was in the face of this realization that Kahneman and Tversky's (1973) studies made such an impact on the field of decision research. As they found, utility theory, (as it is usually interpreted), is grossly inadequate as a model of individual choice behavior. Instead, "People rely on a limited number of heuristic principles, which reduce the complex tasks of assessing probabilities and predicting values to simpler judgmental operations. In general, these heuristics are quite useful, but sometimes they lead to severe and systematic errors" (Tversky, 1975, p. 164).

HEURISTICS AND BIASES

Tversky and Kahneman found that we often use heuristics (mental shortcuts) to inform our judgments. As Kelley (1973, p. 109) argues, the "naive psychologist" in the street uses a method of science that is logically a poor replica of the scientific one—incomplete and subject to bias—but which can proceed on incomplete information, unlike the scientific method. As Tversky and Kahneman noted in their 1973 paper for the Office of Naval Research,

> Most important decisions are based on beliefs concerning the likelihood of uncertain events such as the outcome of an election, the guilt of a defendant, or the future value of the dollar. These beliefs are usually expressed in statements such as "I think that . . .," "chances are . . .," "It is unlikely that . . .". (p. 207)

Also in their 1973 paper, Tversky and Kahneman identified several underlying heuristics that affect judgments. We do not have room here to outline the (many) heuristics that have been identified by psychologists (e.g., Gigerenzer & Goldstein, 1996; Goldstein & Gigerenzer, 2002). But the point to take on with us is that "rationality" is often unreachable because of (1) insufficient information to adequately compare all aspects of a decisions and (2) insufficient cognitive resources to conduct extensive comparisons. Instead, when making decisions involving uncertainty (about outcomes, choices, and consequences), we employ heuristics to lighten the cognitive load and facilitate quick and effective action.

UNCERTAINTY IN THE REAL WORLD

How we deal with uncertainty has always been prominent in the literature on decision-making (Kahneman, Slovic, & Tversky, 1982; March & Olsen, 1976) because in real-world settings uncertainty is a major barrier to decision-making. In the lab studies by Tversky and Kahneman mentioned previously (amongst others), gains were absolute, derived from monetary value with a known level of risk between two options. Yet in the real world, the differentials between options (and especially outcomes) are not so clear, and decision-makers will struggle to determine what has happened; what is happening; and, crucially, what will happen if they act (or do not act).

Uncertainty has had an uncertain history. As Yates and Stone (1992) noted, "If we were to read 10 different articles or books about risk, we should not be surprised to see risk described in 10 different ways" (p. 1). A consensus definition now is "uncertainty in the context of action is a sense of doubt that blocks or delays action" (Lipshitz & Strauss, 1997, p. 150). This definition has three important facets: Uncertainty is subjective (i.e., different individuals may experience uncertainty in different environments and to differing degrees), uncertainty is inclusive (it does not specify a "specific" form of doubt), and uncertainty is directly linked to action (i.e., uncertainty prevents action) (Lipshitz & Strauss, 1997p. 150).

So, questions remained as to how people, in the real world, can still act under uncertainty. Lipshitz and Strauss (1997) sought (from a naturalistic perspective) to understand the strategies people actually employed when dealing with uncertainty using soldiers (members of the Iraqi Defense Forces Command and General Staff College) as their test case. Lipshitz and Strauss asked soldiers to "write a case of decision making under uncertainty from your personal experience in the I.D.F." From this, they were able to determine how soldiers both conceptualized and dealt with uncertainty. The two most common sources of uncertainty were that they inadequately understood the situation (24.6%) and were conflicted due to having alternatives with equally attractive outcomes (24.6%) (p. 156). The most common tactic to reduce uncertainty was to collect more information, seek advice, or apply a standard operating procedure (SOP; a tactic or course action written within doctrine). The second most common tactic was to acknowledge uncertainty by preempting possible outcomes. Finally, soldiers suppressed uncertainty by ignoring it. From this, Lipshitz and Strauss proposed the RAWFS heuristic capturing the five different methods through which a decision-maker navigates uncertainty: (1) *r*educing uncertainty (by collecting more information), (2) making *a*ssumptions, (3) *w*eighing pros and cons of alternatives, (4) *f*orestalling, and (5) *s*uppressing anxiety (Lipshitz, 1997; Lipshitz & Strauss, 1997). These five methods are deployed preferentially in the order displayed previously: Inadequate understanding is reduced by collecting (or attempting to collect) more information and using assumptions to "fill in the gaps." If this is not viable, decision-makers compare courses of action to reduce uncertainty, or they prepared contingencies (forestalling). Finally, if all else fails, decision-makers resort to suppressing anxiety.

NATURALISTIC DECISION-MAKING

As argued by Gary Klein (1989) in his article in *Military Review*, "It was fairly clear how people didn't make decisions." Decision-makers did not, as stated by Klein (2008),

> generate alternative options and compare them on the same set of evaluation dimensions. They did not generate probability and utility estimates for different courses of action and elaborate these into decision trees. Even when they did compare options, they rarely employed systematic evaluation techniques. (p. 456)

This was thought to be especially true in complex environments in which uncertainty is pervasive. Recognizing the disconnect, the United States Army Research Institute funded a conference in 1989 that brought together behavioral scientists who focused on decision-making in the field. The theme of this conference was decision-making under conditions of "time pressure, uncertainty, ill-defined goals, high personal stakes, and the other complexities that characterize decision making in real-world settings" (Lipshitz, Klein, Orasanu, & Salas, 2001). The common view was that although such factors are understandably difficult to replicate in a lab setting, they need to be incorporated into theories of decision-making. Second, the importance of expertise was identified, as was the trend within the wider field to exclude novices when studying high-stakes decision-making. Finally, there was agreement that situation awareness (discussed in Chapter 4) was at least as critical (if not more) than understanding how individuals select between available courses of action. This conference catalyzed interest in studying decision-making in the real world and was quickly followed by a second, third, and fourth annual conference.

Naturalistic decision-making emerged from this conference as a distinct subdiscipline and now has made 30 years' worth of contributions to the field. NDM models have been empirically shown to match what decision-makers do in dynamic and uncertain high-stakes situations (Cannon-Bowers & Bell, 1997; Pascual & Henderson, 1997), and NDM perspectives have been confirmed as relevant to a range of organizational contexts, including aviation, sports, business, engineering, and the military (see Chapter 3; see also Gore, Flin, Stanton, & Wong, 2015). NDM models have become attractive to academics and practitioners seeking to explain decision-making in time-pressured and uncertain situations due to emphases on intuition and expertise. An NDM researcher might conduct field research to discover the strategies people use when making tough decisions with limited time, high uncertainty, high stakes, and unstable conditions rather than beginning with a rational actor model in mind. As Lipshitz et al. (2011) discusses in their outline of the history of NDM, to fulfill this "mission," NDM focuses on five characteristics: proficient decision-makers, process orientation, situation–action matching decision rules, context-bound informal modeling, and empirical-based prescription. The strength of NDM, then, is that it emphasizes the experience and knowledge of the decision-maker (Zsambok, 1997). The "proficient decision-maker" therefore recognizes the role of prior experience on the decision-making process. As such, the decision-maker does not just passively evaluate information in the environment but also uses prior experience to sort incoming details and to direct the search for new information.

Although there are many different NDM models (for an outline of numerous examples, see Lipshitz, 1993), recognition-primed decision-making (RPD; Klein, 1993, 1998) is viewed as the "prototypical" NDM model (Lipshitz et al., 2001). RPD developed serendipitously to explain research on how fire commanders made decisions under time pressure and high uncertainty (Klein, Calderwood, & Macgregor, 1989). The scientists theorized that time constraints would cause commanders to generate a small number of alternatives, falling back between a favored option and an alternative, rather than engaging in the full comparison and analysis of all courses of action. The reality was even more stark: The commanders often carried out the *first* option that they identified. How could commanders rely on the first option they developed? And how could they evaluate this option without considering any alternatives? Klein developed RPD to answer these questions. The key insight of RPD, developing from the idea that a decision-maker sizes up a situation and takes the first course of action, is that the more expertise an individual has, the more feasible that first option will be (Klein, Orasanu, Calderwood, & Zsambok, 1993). When evaluating this first-choice course of action, the decision-maker engages in mental simulation to assess chances of success and to mentally simulate what *could* happen (also referred to as "story-building"; Klein & Crandall, 1995; Pennington & Hastie, 1993). This model is termed *recognition-primed* to imply that decision-makers who have significant prior knowledge in the area are more likely to recognize parallels between current and previous situations and therefore to build *better* mental simulations of what might happen in the present situation. This approach mirrors that of chess masters; rather than laboriously planning many moves ahead, they rely on greater experience with particular board positions to make more accurate characterizations of the strength of the moves available to them. Novices, on the other hand, will tend to try to consciously play out several moves ahead from a candidate move, spending precious cognitive resources to ultimately arrive at a worse course of action. To date, the RPD approach has accurately described decisions made by naval officers, medics, tank platoon leaders, and aviation pilots (for reviews, see Klein, 1998, 2013). In fact, RPD has been integrated into the MDMP to help decision-makers capitalize on intuition and expertise, as well as to speed up military decisions in time-sensitive situations.

THE RECOGNITION PLANNING MODEL OF MILITARY DECISION-MAKING

In 1999, John Schmitt and Gary Klein (1999b) published the results of their naturalistic study of military decision-making. Their presupposition (which we agree with and highlighted previously) was that although the Army, Marines, and Navy have developed formal planning models to assist with the planning of military operations, "these models are inconsistent with the actual strategies of skilled planners, and they slow down the decision cycle" (p. 520). Our interviewees highlighted this. For example, one soldier stated that the people making the decisions on the doctrine have no idea what they're talking about [interviewee laughs]. That's my personal opinion. They do not take into account the feedback they get from the field so that's the reason. And in theory and in a perfect little world and happy bubble we all want to get it all, make it all like roses and fine and dandy but it's not like that.

Schmitt and Klein (1999b, p. 1) argue that "the formal models are usually ignored in practice in order to generate faster tempo." Schmitt and Klein therefore proposed a new model for military operations planning, the recognition planning model (RPM). The RPM, they argued, is consistent with both military planning methods (MDMP) and what we know about human decision-making processes in time-pressured uncertain environments (RPD; Klein, 1998; Schmitt, 1994; Schmitt & Klein, 1996). We view the RPM as both descriptive, in that it is the process that Schmitt and Klein observed planners gravitating toward, and prescriptive, in that it provides a routine that could be followed to increase the pace of the planning process (Schmitt & Klein, 1996, p. 3).

The first goal of the RPM was to streamline the planning process. "Tempo" is the speed of military action and, along with surprise, concentration, and audacity, is a fundamental concept of maintaining the initiative in war (United States Army, 1997, Field Manual, [FM], 5-71-2). The importance of operational "tempo" cannot be underestimated. The Joint Vision 2020, which outlines the US vision for future capability and warfare, specifically states that "faster operations tempos" are a desired capability that will "serve as a catalyst for changes in doctrine, organization, and training" (p. 32). As mentioned previously, the MDMP can often "overrun," decreasing operational tempo and opportunities to maintain (or seize) initiative. Given this, RPM was specifically designed to be compatible with time-constrained planning situations. Second, RPM seeks to ensure that the commander (the most experienced person in the planning unit) is as involved as possible rather than taking a more passive role of approving/disapproving options generated and presented by subordinates. In addition, RPM does not prescribe a linear planning process. Planning, although often sequential, is also dynamic in that activities overlap (especially in large plans) and can move between phases based on feedback. Finally, in line with what Klein (1998) observed in firefighters, the RPM requires decision-makers to make "a tentative decision" early in the process (rather than as an emergent outcome of the process as a whole). This will allow stages of rehearsal, order dissemination, and so on to be planned prior to final approval, increasing time for the implementation stage of the operation. Based on these goals, coupled with their naturalistic observation of mission planning from Commander Joint Task Forces, US Marine

Expeditionary Forces, and US Marines Regiment Combat Operations Centers (among others; Klein, 1998; Klein, Phillips, Klinger, & McCloskey, 1998; Miller, Zsambok, & Klein, 1997), Schmitt and Klein (1999b) proposed a four-stage model of military planning.

The first stage of the RPM is to "identify the mission/conceptualize a course of action." Contrary to the MDMP, the RPM acknowledges that planners actually simultaneously identify the mission and conceptualize a series of rough options. This is in fact beneficial for the decision-maker because conceptualizing a viable option also helps clarify the mission as "planners gain a better understanding of the problem by posing and then working through potential solutions" (Schmitt & Klein, 1997, p. 6). The RPM also puts far less weight on identifying and comparing courses of action based on the findings (as outlined previously) that decision-makers do not compare multiple courses of action in parallel but instead accept and reject courses of action in sequence until they find one they think will work. As Schmitt and Klein argue, "Planning procedures that require the creation of multiple courses of action are artificial and, in our observation, do not improve the quality of plans" (p. 7). The outcome of this stage of the RPM is to "identify" a decision—namely a tentative concept of operations. This early decision is important because it increases operational tempo while increasing the amount of time that can be taken to arrange certain aspects (e.g., mobilizing air support). In Schmitt and Klein's view, "A good concept well-executed is superior to a superior concept poorly executed" (p. 6; clearly a nod to George S. Patton's adage that "a good plan violently executed now is better than a perfect plan executed next week"). If anything, this shows that the RPM is rooted in pragmatism and seeks to both reflect and support the *actual* requirements of decision-makers.

The second stage of the RPM is to analyze and operationalize a course of action. Planners cannot conceptualize a reasonable option without knowing if it is feasible or how it will be executed. This is therefore the point at which a plan meets reality, and doing so early is critical. For example, if a central requirement of a course of action is support from an allied force, then it is better to determine if that force is willing and able to provide support as soon as possible. After it has been confirmed that an option is indeed "feasible," it is war-gamed (similar to the MDMP). Klein and Crandall (1995), along with many others (e.g., Kahneman & Tversky, 1982; Klein, 1993), argue that mentally simulating the outcomes of an action is vital for effective decision-making. The role of mental simulation was also emphasized by our interviewees in that, by mentally calculating the possible outcomes, they saved themselves from paralysis when these situations occurred during execution. The following was recalled by one of our interviewees about how he used mental simulation to plan operations in Afghanistan:

I call it playing chess, but I try to visualize the chess games, like I do this, what is the reaction? If I do this, what is the reaction? If I do this, what is the worst that can happen? And when I'm [facing these decisions] I am not doing it by myself obviously, I have some field grade officers and helping out with the decision and saying what about this and what about this, and we can do this. And by kind of thinking through that we'll red team it. Try to figure out what is the most likely thing to happen that we had to be prepared for and what is the worst thing that can happen and are we prepared to handle that? And we kind of walk through each of those scenarios, and then just make sure we have some kind of counter-measure to, so, if it is likely that somebody put an IED [improvised explosive device] on the route between here and there, then let's make sure we have route clearance between here and there. Ok, we are likely to be on the ground for 8 hours, ok, well we could become a sitting duck for a sniper so let's put an extra cordon on there. And, oh yeah, 8 hours, you guys are going to need a latrine to drink water and maybe something to eat because it is going to take us 3 hours to get there and 3 hours back so let's think about some logistics of that stuff. So yeah, so, did I put a mental framework on that, yeah, and I think that started back in the early tours where you try to visualize how this unfolds, you know, . . . what is the most likely thing that is going to happen, and then um, I think the last thing, at least as a leader is, the last thing you visualize, is what if the worst does happen? What am I going to do about that, what do I have in reserve, who do I call? What if this just goes completely south? What does that mean to me? What are the first 2 or 3 things that I am going to do, because there is not an SOP for that. But it is nice to have thought about that, because then you do not suffer from the paralysis.

Whereas the traditional MDMP involved war-gaming all possible courses of action (as a part of the evaluation of feasibility), the RPM envisions war-gaming a single course of action against several *enemy* courses of action. This means that more time can be invested in anticipating how to deal with the consequences of an action rather than just its implementation. This focus on predicting the enemy's reaction to a course of action is especially important given that, as argued by military strategist von Moltke (1800–1891), "You will usually find that the enemy has three courses open to him, and of these he will adopt the fourth." Finally, once the course of action has been war-gamed, approved, and become a plan, the necessary order documents are developed (although usually this will have started far earlier in the process).

In comparison to doctrinal approaches such as the MDMP, the RPM offers several advantages. First and foremost, although it is sequential, it is not linear, meaning that stages can overlap and co-occur. This means that the final stage of the RPM (order generation) has likely been ongoing since the end of stage 1. Furthermore, the feasibility of a course of action is assessed much earlier, meaning that time and energy are not invested in making a decision that, in reality, is not a decision. The importance of this cannot be understated. Consider the following decision faced by General Petraeus while he was commander of the International Security Assistance Force (ISAF) and Commander, US Forces Afghanistan (USFOR-A; cited in Atkinson, 2005; see also Shortland & Alison, 2015):

> One seemingly trivial item on Sinclair's agenda was in fact vital: Should helicopter blades be taped or painted? Apache and Blackhawk rotors revolve at such high speeds—1,456 feet per second at the tips—that blowing grit could bore through the titanium spar on the leading edge of each blade. Wormlike, a grain of sand would then eat out the honeycombed material inside the blade, which might unbalance the helicopter aerodynamically and cause a crash. Traditionally, the blade edges were protected with strips of black tape, which had to be reapplied after every mission or two. But taping was time consuming, difficult in the desert, and required adhesive that wore badly in hot weather. Some aviation experts insisted a thick coat of black paint, reapplied to the edges after every flight, was an effective substitute.
>
> Rotor blades were in short supply—the 101st had only five spare Blackhawk blades, which cost $80,000 each. More to the point, each Apache cost $20 million, and each Blackhawk carried the souls of four crewmen and as many as sixteen passengers. The tape-versus-paint conundrum neatly illuminated the thousand technical challenges facing every commander. . . . The issue had been hotly debated between the commanders for months, with some aviators stating "I'll go to my grave before I put tape on." General Petraeus remained torn, he had heard some good things about tape but in his gut he remembered the problems tape had caused in the past. The issue however was moot. Hardly a single roll of tape for the 101st could be found in Kuwait. The division's supply had been stored in an East Coast warehouse that had collapsed during a recent blizzard. All that discussion . . . And there's no tape. (pp. 54–57)

This example shows a clear strength of the RPM: The quicker one can begin to operationalize a course of action, the quicker one can determine if it is actually feasible. In RPM, these tests of feasibility are done much sooner, increasing operational tempo and lowering the risk that resources and time will be invested in an unrealistic course of action.

CAN THE RPM REPLACE THE MDMP?

The RPM has been experimented with by several military commands, including the US Marine Corps and the British Military (Pascual, Blendell, Molloy, Catchpole, & Henderson, 2004). In addition, Peter Thunholm, a psychologist with the Swedish Defense University, compared division-level planning groups within the Swedish Army that used either the doctrinal MDMP or the RPM, finding that RPM increased operations tempo by 20% (Thunholm, 2005). Because the RPM did not seek to replace the MDMP, but merely codify the decision-making process, it was well received, as the military personnel commented they were "already doing this" (Ross, Klein, Thunholm, Schmitt, & Baxter, 2004).

Ross and colleagues (2004) tested the RPM against the traditional MDMP during a 2-week experiment at the Fort Leavenworth Battle Command Battle Laboratory (BCBL). Two days were devoted to training staff on the RPM. The scenarios used across 5 days of testing involved multiple planning loops and variations of offensive operations. The staff also had to plan stability and support operations after the offensive action was over. The decision-making of the teams was observed by researchers, who administered questionnaires and performed in-depth interviews. The results were mixed. Participants had little trouble learning and using the RPM (unsurprising because it is designed to reflect what they were already doing), and they estimated that it took 30% less time than the MDMP. However, some participants stated that the RPM caused them to rush through mission analysis (which in RPM occurs in tandem with course of action analysis). Furthermore, despite using RPM, participants gravitated toward several MDMP tools (e.g., listing the assumptions they were making). One participant (a colonel) cautioned throwing away 26 years of the MDMP because of 5 days with the RPM. Although he believed that the RPM had demonstrated sufficient face validity to warrant additional research, this demonstration alone was not sufficient to justify replacing the MDMP (Ross et al., 2004, p. 10).

One important flaw in RPM was raised by participants: the degree to which the RPM helped them make decisions in novel environments (Ross et al., 2004). Clearly, if a commander lacks experience (central to the RPM), then he or she will generate lower quality plans when using the RPM. This is a valid concern given that a central objective of the RPM is to let the commander's experience increasingly drive the planning process. However, this criticism is not unique to RPM; it also affects the MDMP. In fact, new situations in general cause major problems for the RPD approach. Let us examine the original model of RPD (Klein, 1993, 1998; Figure 2.1).

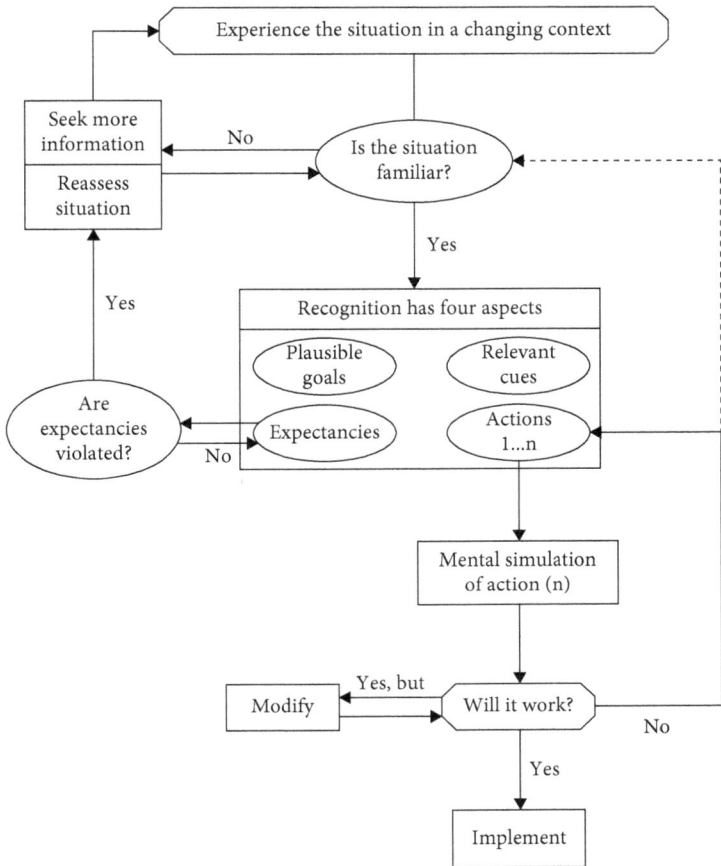

FIGURE 2.1 A model of recognition-primed decision-making.
Source: Reproduced from Klein (1993).

RPD/RPM AND LEAST-WORST DECISIONS

Both RPD and RPM rely heavily on the experience of a leader or commander, specifically as it pertains to their ability to pattern match the current situation to a familiar one (Klein, 1997). The issue, then, is that given the diversity and complexity of the contemporary conflict environment, commanders (as well as more junior officers) can often face situations for which they have no "analogous" experience. This is an irrefutable issue.

In such circumstances, how do decision-makers adapt and improvise? And what degree (if any) of transferability of expertise is there from one type of military domain to another? In conducting our own research, we noticed this becoming increasingly enmeshed with least-worst decisions. The more members of the Armed Forces we interviewed, the more often we heard least-worst decisions being described as something they had "never experienced before," in which they had no "analogies" they could apply or training that could help them decide what to do. One interview, for example, was with a driver working in Afghanistan and providing rear support to a large convoy of oil tankers that needed to be delivered to a forward operating base. During one trip, he had to use his truck to protect the convoy from what he thought could have been a vehicle-borne IED that was heading directly for them. He explained this situation as unique:

> This particulate situation was unique, and, especially where we had two things happen one after another, I mean, this is 2010. This is when the whole drawdown was going on and everything . . . these were pretty quiet times. . . . But yeah this was really a unique situation, I cannot think of another situation where it was like this. . . . So, yeah. I mean, we'd had trucks try to force their way into our convoys but never like this. You know, one going 70 miles an hour, just coming off the on-ramp thing and he is just going to pull right in, like no big deal. Not like this. I'd have to block vehicles, or I'd have to force vehicles out of the way, you know. Um, I mean there was many many times I'd have to use the truck to my advantage, but not to the point where I thought I was blocking a potential IED, it was more keeping the vehicles out. Keeping our convoy clean so we could keen eyes on and keep protection of the people the trucks were trying to protect.

Another interviewee, a drone pilot, shared a similar sentiment:

Interviewer: So how did this decision differ to the ones you'd faced in training?
AFC Evans: Training is easy, you know you just do it you're given missions without thought and you're not giving, you're never given scenarios where Afghans may be doing this or they may not. Go look at this, find this, do this, get a beer after or a steak dinner. It's not like this where you live on a base that's getting rocket attacked and stuff. Everything plays into it. There were a few times we were wearing our gear while we were flying because we were getting attacked. Nothing training can do.

This is one reason least-worst decisions are so hard, and they fall in a gap that RPD cannot satisfactorily explain. What happens when a situation does not match a previous one and there is no pattern to match? As shown in the model in Figure 2.1, when facing situations that have no analogy, we seek more information. But extensive information searches are often impossible due to lack of time and lack of available intelligence.

This is a critical point. If we look more closely at the types of decisions that people have made when they use RPD, they are (to varying degrees) similar. In military cases, RPD is often dominant when planners are dealing with common, well-rehearsed missions: Thunholm's (2005) research involved defending the troops from an enemy ambush. Although such cases are the norm for military decision-makers, often the norm is upended and a soldier must choose an option in a situation that he or she has never encountered.

CONCLUSION

Least-worst decisions parallel Taleb's "black swans" (i.e., highly improbable but incredibly damaging situations, such as the stock market crash of 2008; Taleb, 2008; see also Posner, 2010). These kinds of situations have the highest risk of a negative outcome, yet the critical factor—and what makes them so hard—is that they are novel and we have no guiding experience to reach for. It is therefore crucial to understand how decision-makers both thrive and falter under these circumstances, which constitutes an important gap in extant models of decision-making. To fill this gap, we introduce a phase-based model, built from our observations with experienced practitioners, that we then use to examine and explain the process through which soldiers make the least-worst decisions they face.

3

THE SCIENCE OF SELECTING LEAST-WORST OPTIONS

"Oops"? What the fuck do you mean "Oops"?
—Unidentified camera operator to a Predator drone pilot before takeoff, July 24, 2012

Military decisions have zero margin for error. "Oops" as a response to breaking an egg for making an omelet is acceptable, but the consequences of implementing decisions where outcomes mean life or death cannot be so casually considered. In this chapter, we present the SAFE-T model of decision-making—a descriptive model we hope offers an innovative insight-generating platform through which the process of making military decisions can come into better focus. Why? Because it is simpler than most models, and thus quicker, pragmatically connected to the military decision-making process (MDMP) in terms of being phase-based, and supported by empirical evidence (as well as retaining components of experience and expertise). There are several more specific benefits to applying the SAFE-T model to military decision-making. First, the phased nature of the SAFE-T model helps identify not only factors that influence the decision-making process but also how the *current phase* is influenced differentially by each factor. The SAFE-T model is principally focused on the ways in which decision-making can become "derailed." As such, it identifies not only how team-based factors (arising from the people responsible for managing and responding to a situation) and situation-specific factors (e.g., uncertainty and time pressure) in the environment affect the decision-making process but also when in the decision-making process these factors exert the most influence. The SAFE-T model is closely aligned with "indecision" (or "doing nothing"; Anderson, 2003) and, as such, shows where things can go wrong as well as where they should go right. The SAFE-T model also identifies how derailment can occur throughout the decision-making process (i.e., "the errors on the way")—namely decision avoidance (postponing attempting to make a decision), decision inertia (being unable to decide on a course of action), and implementation failure (failing to put a decided upon course of action into action).

THE SAFE-T MODEL OF STRATEGICALLY CHALLENGING DECISIONS

The SAFE-T model was developed by van den Heuvel, Alison, and Crego (2012) and was based on extensive evaluation of the strategic decision-making literature and a detailed naturalistic observation of decision-making in critical incidents (Alison et al., 2013). The SAFE-T model identifies four key phases in accurate decisions: situational assessment (SA), plan formulation ([P]F), plan execution ([P]E), and team learning (T[L]). Consider, for example, that you are walking along a quiet street and you see an altercation between two people. First, it is important that you understand what is going on (SA). Are they fighting? If so, is it a fight or a mugging? Who is the aggressor? Once you have a good (or the best possible) SA, you need to work out what you can do (F; which in this case includes intervening, calling for assistance, phoning the police, or doing nothing). You then need to choose a plan and, crucially, actually do it (E). Once you have acted upon the situation, and the event has unfolded, you then reflect on what happened, what you chose to do, and if your choice led to a good outcome (or not; T). These stages reflect the general phase-based process that most individuals go through when making decisions (critical or otherwise; e.g., Lipshitz & Bar-Ilan, 1996).

In Figure 3.1, we overlay the SAFE-T model on the MDMP to show their similarities. Situational assessment is the process through which an individual comes to understand his or her environment by identifying and encoding salient cues present within it. This also allows the individual to develop expectations about what might occur (Klein, 1993). In military terms, situation awareness is "knowledge and understanding of the current situation that promotes timely, relevant and accurate assessment of friendly, competitive and other operations within the battle space in order to facilitate decision making" (Army Field Manual 1-02, 2004). Within the MDMP, PF most closely represents course of action (CoA) development and analysis. PE relates to the point at which a tactical decision has been made, and these plans are then implemented. This is analogous to the process of CoA approval in the MDMP. TL applies to those situations that are "slow burn," meaning they are multiphased, and within which the decision-maker has the opportunity to implement a decision, learn from the outcome, and self-correct before launching a new decision (Salas, Rosen, Burke, Goodwin, & Fiore, 2006). In such cases, decision-makers are able to continually reflect on and revise assessments throughout the process, allowing them to adapt future responses to fit the demands of a dynamic and volatile situation. This is often facilitated by feedback from team members and the outcomes of decisions that have been implemented (House, Power, & Alison, 2014).

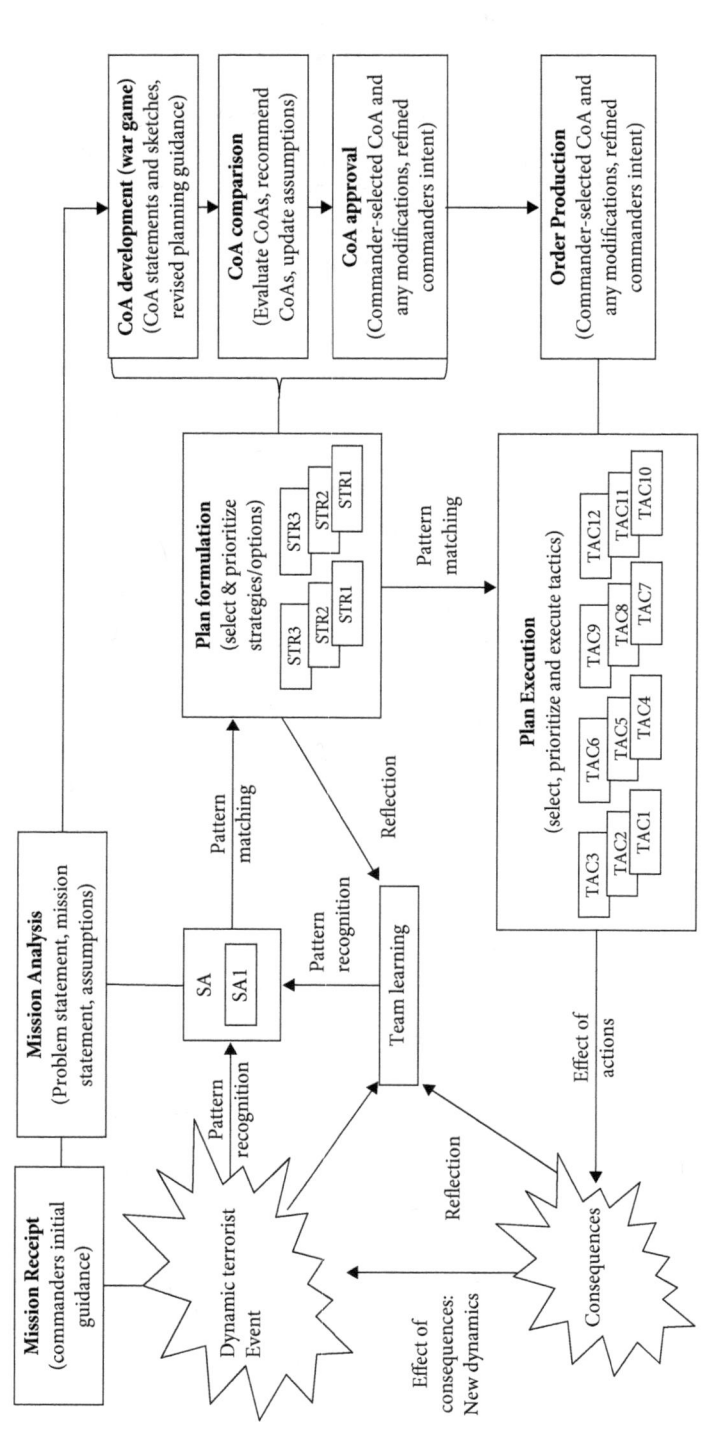

FIGURE 3.1 The SAFE-T model of military decision-making shown in reference to the MDMP.
Source: Neil Shortland.

DECISION INERTIA AND "FAILURES TO ACT"

A central strength of the SAFE-T model is that unlike both recognition-primed decision-making (RPD) and the MDMP, it accounts for the phenomenon of decision inertia. Decision inertia is a phenomenon that Alison and colleagues have found consistently in critical incidents. In Chapter 1, we used a common definition of a decision as a "commitment to a course of action." "Action" is therefore central to decision-making. If the decision-maker has not operated on the world, this is indistinguishable from when a decision has not been made. In the previous example about encountering an altercation, even if intervening is the *choice* you made, it does not become a *decision* until you have actually engaged in the *behavior*. Decision inertia is therefore a failure to act (Power & Alison, 2017). Given that military plans must be executed in a speedy manner, theories that focus on potential points for indecision should be especially pertinent. Previously, when we discussed what makes a decision challenging, we centered on "decisional conflict," which occurs when there are "simultaneous opposing tendencies within the individual to accept or reject a given course of action," and the consequences of which involve "hesitation, vacillation, feelings of uncertainty, and signs of acute emotional stress" (Janis & Mann, 1979, p. 46). Thus, although naturalistic decision-making (NDM) and RPD have attempted to understand how humans make fast and effective decisions in complex real-world settings, such as firefighting (Klein, 2008, p. 16), they have overlooked the process through which decisions are *not made* in complex situations.

Before we continue, it is important to highlight that we focus on decision inertia over the more commonly studied phenomena of indecision. Consider, for example, the following dilemma:

> Consider the option of inoculating a child against a respiratory illness. There is a reasonable chance that the vaccination will result in a heart condition. However, a failure to inoculate could lead to the respiratory illness and death.

This dilemma presents two equally adverse (or beneficial) options and is likely to result in indecision. This indecision can take one of several forms. First, the dilemma can be avoided ("I will think about this next year or when school finishes"). This is a common form of indecision (Anderson, 2003). Second, the decision-maker can engage in the cognitively heavy process of trying to decide between the options, weighing the benefits and costs of each and forecasting potential future outcomes. Although at the start this reflects a rational–cognitive approach to decision-making, once this cognitive rumination continues for no further gain (i.e., the decision-maker is no longer making progress but continues to ruminate), we refer to this active process as "decision inertia." Finally, the decision-maker can *decide* (i.e., commit to a choice of action) but fail to implement the choice. These three forms of indecision represent distinct, mutually exclusive psychological processes. However, not all of these forms of indecision are equally relevant (or important) to military operations and military decision-making. Because most military operations are time-sensitive (and initiated by higher chains of command), decision avoidance is impossible. Nor, for the same reasons, can an individual decide and fail to act (in the same sense that a career or purchasing decision can be derailed). However, it is clear that military decision-makers can and do struggle to evaluate equally adverse options. As one of our interviewees described it, soldiers can suffer from "paralysis by analysis"—also known as decision inertia.

When facing uncertain or equally adverse options, an individual is likely to delay the choice, and psychological research has identified both the personal antecedents and decision-related aspects that result in indecision. Unfortunately, all forms of indecision are often lumped together as a "failure to act" and appear to be viewed as inextricable from procrastination (Tibbett & Ferrari, 2015). The current literature on indecision is therefore insufficient to support our understanding of military failures to act. Decision inertia is a cognitively busy process of continued redundant deliberations about options and forecasting of future consequences. Decision inertia involves a *motivated* decision-maker who, despite considerable mental effort, is unable to make a decision, either in time or at all (van den Heuvel et al., 2012). Thus, while "non-decisions" have been considered a static outcome, our approach provides increased depth to explain the ability of individuals (both military and non-military) to make timely and effective decisions when under conditions of risk and uncertainty.

It is worth laboring for a moment over the contrast between inertia and decision avoidance. Avoidance, as Anderson (2003) neatly defines, is "a tendency to avoid making a choice by postponing it or by seeking an easy way out that involves no action or no change" (p. 139). Motivational drives underpin decision-making. Motivation is central to emotion regulation (Elliot, Eder, & Harmon-Jones, 2013)—it energizes, then translates thoughts into actions (or, here, inactions). Avoidance, then, involves an individual wishing to not approach a stimulus (e.g., decision) because it is associated with anticipated negative affect (Elliot, 2006) ("If I do that, it's going to turn out really bad and I will feel terrible, so I'm going to avoid doing anything about it"). Anderson describes the several ways in which decisions may be avoided:

- Status quo (i.e., "I will go along with the majority.")
- Choice deferral (i.e., "I will postpone my choice and wait and see what happens.")
- Omission (i.e., "I will not think about this choice.")

When decisions are more difficult to avoid (e.g., because of a deadline), negative emotions increase and choice is perceived as more complex (Mamhidir, Kihlgren, & Sorlie, 2007). However, inertia is rather different because the decision-maker does not avoid thinking about the problem but, rather, does not *stop* thinking about it (at least in terms of options). Thus, as with ruminating over whether to intervene in Syria (and, if so, how much and in what way), as well as what to do about the refugee crisis, these examples are not decision avoidance but, rather, decision inertia—there is plenty of thinking going on, but there is a lack of actual well-defined strategic end-goal planning. Inertia is effortful but ultimately unproductive thinking and often appears as redundant deliberation over options (Eyre, Alison, McLean, & Crego, 2008), where present outcomes, such as which option is best or which is least-worst (Parker & Schrift, 2011), and anticipated potential negative future outcomes (Beeler & Hunton, 1997), such as being held to account for making a wrong decision (Mamhidir et al., 2007), are weighed and reweighed repeatedly in the decision-maker's mind.

In summary, we define inertia as the redundant cognitive deliberation over choice for no positive gain (Alison et al., 2015; Power, 2016). Paradoxically, cognitive processing does not help the decision-maker in reaching a choice because there is no new information to help tip the scales. The decision-maker is inert—effectively he or she is trapped on a cycle of continual re-evaluation to try to trade-off salient competing options, goals, and anticipated potential consequences. But inertia is *not* a function of wishing to disengage and avoid a choice; rather, it is as a function of wanting to avoid *loss*. It is not about energizing behavior toward maximizing the most positive outcomes but instead reflects internal cognitive dissonance as individuals try to accept the least-worst negative outcome. Given this, we propose that decision inertia is not only a subset of wider theories of indecision—separate from avoidance and implementation failure—but also the most important (and potentially harmful) form of indecision in military environments. Thus far, no academic research has explored decision inertia within the military decision-making process. However, our decades of work observing (and training) practitioners experiencing decision conflict has long-confirmed that they are prone to indecision. Let us consider a decision that we often pose to delegates (usually police and fire officers but sometimes military personnel and students) in training events. This example often results in decision inertia.

THE TUNNEL SCENARIO

You are a fire officer with responsibility for tactical level interventions in the following situation and you are "on-scene" silver commander. The time is now, and you have responded with upwards of 10 staff to a tunnel collapse that appears to have occurred as a function of an explosive device midway in the Birkenhead Tunnel [you can use any long tunnel in a city you know well]. There is no fire, but significant collapse, metal and concrete displacement, and there are at least 30 casualties. At present, it is not known if anyone has died. All your staff are midway in the tunnel and are busy working on casualties and triage with the ambulance service. You then receive the following information from a police officer, who tells you that there is "credible information there is another IED [improvised explosive device] in the tunnel which is likely to go off in the next 15 minutes and has been claimed by a terrorist organization with the explicit intention of killing emergency service staff." What do you do?

This generates a number of rapid-fire questions from the audience, but rather quickly most will agree to pulling staff out of the tunnel to avoid further deaths or casualties. So, in terms of decision and action, response to this is fairly swift. Thus, even though the consequences of all actions are high, this is not a difficult decision. However, what follows *is* a very difficult decision:

You have now actioned the evacuation of responding staff. However, 4 minutes into this process you receive the following message from a fire officer: "It's Mark Cunningham here. . . . I've got an 8-year-old girl in here—she is trapped by a metal girder. I need pedal cutters. Get me some pedal cutters and I can cut her out in a couple of minutes. I'm not going to come out and I'm not going to leave her until I get some pedal cutters." Now what do you do?

What many of our participants will do is ask several more questions: "How far in the tunnel?" "Is the girl conscious?" "How credible is the intelligence?" "Have I got people volunteering?" "Do those volunteers have children of their own?" and so on. That is, they take time to explore more options—sometimes to the point of adding another 20 minutes to the scenario. When participants are spending their time asking more and more questions, to the point where either there is no more useful information to learn or the bomb goes "bang," they are not actually making a decision (i.e., they have not refused to go in, gone in themselves, or sought volunteers—they have just kept asking questions). Needless to say, asking for more information to make a better assessment makes sense—usually. But, if there isn't time, someone must take charge—decision and action need to occur. Effective decision making is therefore also about *when* to decide. Thus, it is fine and indeed good thinking to ask further questions, but when the answer keeps coming back as "I don't know" or "I don't have any other information," the person in charge needs to assess the situation, formulate a response, and act. This, perhaps, is what best separates decision inertia from wider failures to act (and from research on indecision in other areas of psychology, such as consumer purchases or career decisions; Miller & Rottinghaus, 2014) in that it is about engaging in a prolonged cognitive deliberation—seeking unavailable information—in a situation in which a decision needs to be made.

As naturalistic researchers, we have observed live decision-making in a range of events from responding to a counter-terrorism event to simulated responses to fires, and we have identified several broadly chronological stages at which one can fail to act:

- A failure to start thinking about the problem (ignoring): Saying that you are "not going to worry about Mark Cunningham and that girl, I've got bigger/other things to think about."
- Starting but then stopping (disengaging/avoiding): Feeling that "Christ, this is difficult. . . . I don't know what to do. . . . I need to leave this to someone else/do something else."
- Thinking but not deciding (inertia): Worrying that "If I go in I could die, but if I don't they could both die, but if I send someone in they could all die, perhaps I should insist he comes out. Maybe there is some other option I haven't thought of."
- Deciding but not acting (implementation failure): Deciding that you are going to use a volunteer—but then waiting until you finally send one in.

Based on our work with emergency services and law enforcement in scenarios such as this, observation of real critical incidents, public review, our own quasi-experimental research, and our operational debriefs, inertia is by far the most common process and—interestingly—it involves the most thinking. Thus, rather than reducing cognitive load (as in other, adaptive, decision-making processes such as applying heuristics), decision inertia increases cognitive load. It also has the potential to be extremely damaging. For example, the 2004 Boxing Day tsunami resulted in 220,000 fatalities, and there were significant delays in delivering aid to affected communities (Rencoret, Stoddard, Haver, Taylor, & Harvey, 2010). The response to Hurricane Katrina in 2006 received similar criticism, in that "bureaucratic inertia was causing death, injury, and suffering" and "indecision about evacuations led to delays" (Select House Committee, House Report, 2006, p. 1). Thus, although decision inertia is an interesting psychological conundrum that gives us insight into human behavior, more importantly it often leads to significant human suffering.

SAFE-T MODEL AND DECISION INERTIA

The SAFE-T model holds that during a critical incident, effective decision-makers will follow a process of situational assessment, plan formulation, plan execution, and iterative team learning. However, our model shows that there are a host of ambient, affective, cognitive, and organizational factors that can derail a decision-maker's ability to navigate this sequence. Furthermore, because decision-making in such events is dynamic, decision-makers can also skip between stages (similar to the feedback loops in the recognition planning model of military decision-making proposed by Schmitt and Klein [1999b]). For example, van den Heuvel et al. (2012), who's research involved a simulated counter-terrorism operation that unfolded during a large-scale public event, found that

> generally, this predicted SAFE-T model was not employed by teams managing a simulated counter-terrorism operation. Specifically, the predicted phases were not always followed in the expected sequential manner, with groups either skipping a decision phase entirely (SA to PE), or making decisions in reverse order (PE to PF). (p. 181)

However, they also reported that "teams were not implementing critical decisions that they should have made in a timely manner, potentially culminating in detrimental consequences for the investigation" (p. 181). Instead, van den Heuvel et al.'s study (confirmed since by naturalistic research; e.g., Alison et al., 2015; van den Heuvel, Alison, & Power, 2014) found that effective decisions were delayed through either "non-actions" (not making a decision) or *apparent* actions (deferring choices or passing issues onto other agencies). These delay tactics constitute derailment from the SAFE-T decision-making process.

FACTORS THAT AFFECT THE ABILITY TO MAKE STRATEGICALLY CHALLENGING DECISIONS

The SAFE-T model (along with other rational models such as that by Anderson [2003]) identifies multiple factors that can stall decision-making at specific stages in the process, and in the following chapters we explore how such factors affect specific phases of the decision-making process. However, in addition to phase-specific factors, there are also several pervasive (and often detrimental) influences on all phases of the decision-making process. Here, we briefly outline each of these.

Regret

Anticipatory regret involves forecasting that negative affect will be experienced in the future based on decisions or actions that could be taken in the present (Wong & Kwong, 2007). In some situations, such as in the case of those contemplating suicide (Lester & Brockopp, 1973; Litman, 1971), anticipatory regret can be highly beneficial in that it prevents us from seizing seemingly attractive options without forethought of the consequences (Janis & Mann, 1977). However, in some cases, anticipatory regret can also cause indecision because the decision becomes preoccupied with hidden risks. In addition to worrying about how we might feel given a certain action, we are often highly attuned to what we might lose. Anticipatory regret is experienced by decision-makers when they are aware of the cost of choosing the most attractive option, especially when the losses are imminent, there is a degree of social commitment, they are optimistic that a better solution could be found (in time), and they are *not* under significant time pressure (Janis & Mann, 1977). An awareness of these costs is especially pertinent when the situation is perceived as "undo-able." Being able to undo actions and return to the status quo is referred to as "outcome mutability," and the likelihood that an individual will experience anticipatory regret is closely linked to this (Morris & Moore, 2000). Mutability is also strongly linked to the "lost opportunity" hypothesis (Beike, Markman, & Karadogan, 2009), whereby decision-makers avoid launching interventions that offer a low chance of being "fixed" if the outcome is bad (e.g., choosing to fire a weapon), because doing so will preclude opportunities that would have been available to them if they had taken an alternative (or no) action. When experiencing anticipatory regret, decision-makers may therefore prefer less risk ("better safe than sorry") or more risk ("better risky than regretful") (Inman & Zeelenberg, 1998; Zeelenberg, 1999; Zeelenberg & Beattie, 1997; Zeelenberg, van den Bos, van Dijk, & Pieters, 2002). Anticipatory regret is also closely linked with decision avoidance, whereby decision-makers select courses of action that maintain (as best as possible) the status quo, minimizing anticipated regret but potentially resulting in inappropriate or zero action (Anderson, 2003).

The issue with anticipatory regret is often that we are quite poor at predicting how we will actually *feel* in a given situation (Gilbert, Driver-Linn, & Wilson, 2002; Gilbert & Wilson. 2000; Kahneman, 1994; Kahneman & Snell, 1990; Loewenstein & Frederick, 1997; Loewenstein & Prelec, 1993; Loewenstein & Schkade, 1999). Woodzicka and LaFrance (2001), for example, asked women to predict how they would feel in response to a sexually harassing question during a job interview. The majority predicted being angry and only slightly afraid. However, when these same women were actually interviewed for a job and asked a sexually harassing question, their predominant reaction was fear, and relatively few reported experiencing anger.

We also overestimate the duration of our future emotions (the "durability bias"; Gilbert et al., 1998). For example, Wilson, Wheatley, Meyers, Gilbert, and Axsom (2000) found that college sports fans overestimated how long they would be happy for the day after their team won a football game. This is closely related to "impact bias"—the tendency to overestimate the impact future events will have on our emotional reactions (Gilbert et al., 2002). Thus, in highly complex and often emotive situations, our actions can be stalled (temporarily or, in some cases, entirely) due to fears about what will happen and how we will react. One of our interviewees shows how the pervasiveness of anticipatory regret affected his decision-making:

> But there is the other side of me that looks at it and says, you know, that if I fired and wasn't supposed to, not only would that have probably ended my career, forget my career, I would have had to have lived the rest of my life knowing I had killed that guy. I don't know what other people's impression of the military is, if we take these decisions lightly, but I certainly didn't and I don't think other people do either. And the idea that you become jaded to the point when you stop caring about hurting innocent people. I can't imagine becoming that jaded about that decision, in fact it still bothers me to this day, the idea that I, in a fraction of a second, I could have shot that guy, and if he was you know, Taliban, great, but if he wasn't, you know, I can't imagine the thoughts that would go through my head about that guy's family, and things like that. So, it is a hard decision to live with.

Accountability

Anticipatory regret is linked to the subjective feeling of accountability. Accountability is a central aspect of all decision-making, and it was commonly referred to throughout our interviews with soldiers, even those who were not the highest in the chain of command (and hence could have deferred responsibility):

> So, if you have a car-bomb that goes off in close proximity, or there is someone that blows a building up and it goes crashing down, uh, then as a commander I have responsibility to find all the personnel, find all the sensitive items, get everything back out of there. And um, so, yeah, that was, all that stuff is important in all three sentences. And then there was, at the end of it, there was the accountability that I was assuming with the decisions we make. Because at the end of it my boss was in Kandahar, probably a two-day trip away from me. Uh, he was trying to be a good boss over the phone, and he [was] trusting my input. At the end of the day, having watched this thing unfold he was

essentially taking my recommendation so I was accountable to him. In the end, I would have certainty been held accountable.

Accountability involves being aware of appraisal by external audiences who have the power to instigate rewards or punishments (Baucus & Beck-Dudley, 2005; Klehe, Anderson, & Hoefnagels, 2007) and also the requirement to provide a justification to these audiences (de Kwaadsteniet, van Dijk, Wit, De Cremer, & Rooij, 2007). This constrains actions because decision-makers know that they will be identified, judged, and punished (or rewarded). It is therefore the psychological antithesis of anonymization (the process of being unidentifiable) and, logically, has the reverse effect. As is well known by anyone who studies (or has ever used) the Internet that anonymity increases hostile behavior (Zimbardo, 1969). Accountability, on the other hand, constraints behavior.

Accountability can lead to behaviors both adaptive and maladaptive; therefore, unsurprisingly, the findings are mixed as to whether accountability improves or degrades decision-making (Hall, Bowen, Ferris, Royle, & Fitzgibbons, 2007). One line of research that provides good evidence that accountability can hinder performance is on "choking" in sports. Beilock, Carr, MacMahon, and Starkes (2002) studied the performance of golfers (professionals and novices) and found that increased inward-focused attention (which can be caused by accountability) can be counterproductive. Specifically,

> Attention to high-level skills results in their "breakdown," in which the compiled real-time control structure of a skill is broken down into a sequence of smaller, separate, independent units—similar to how performance may have been organized early in learning. Once broken down, each unit must be activated and run separately, which slows performance and, at each transition between units, creates an opportunity for error that was not present in the "chunked" control structure. (p. 8)

Ashcraft and Kirk (2001; see also Eysenck & Keane, 1990) suggested anxiety generates intrusive worries that occupy working memory capacity. Beilock et al.'s work therefore suggests that accountability (if it causes anxiety in the decision-maker) could negatively affect sports performance by occupying resources that should be applied to the problem at hand. Similarly, least-worst decisions in military contexts require significant cognitive capacity (working memory and executive functions), and that capacity can be reduced by focusing on external pressures such as accountability.

Naturalistic research provides evidence that the subjective experience of accountability affects decision-making. For example, Waring, Alison, Cunningham, and

Whitfield (2013) found that in a simulated hostage negotiation crisis, accountability increased the motivation for self-preservation and detracted the decision-makers' attention away from the task at hand. This inhibited the decision-makers' ability to discriminate between critically relevant and irrelevant information (Waring et al., 2013). The SAFE-T model explains how accountability will affect the decision-maker during different stages of the decision process. Anticipation of accountability can affect information gathering and interpretation processes during the SA phase. Accountability can also encourage a decision-maker to consider more information without first discerning its relevance, increasing cognitive load for little gain as the decision-maker sifts through data that may not aid in the decision (Tetlock & Boettger, 1989). In the PF and PE phases of decision-making, accountability encourages a decision-maker to switch toward egocentric, defensive justifications (Gollwitzer & Moskowitz, 1996). Police officers facing a dynamic terrorist event shifted priorities away from saving the lives of those in a (potential) attack location to saving themselves (i.e., making decisions that could be defended if later reviewed [van den Heuvel et al., 2012]). Similar to anticipatory regret, accountability is also linked to inaction. This is entirely consistent with the "omission bias" (Spranca, Minsk, & Baron, 1991) in which harmful commissions (i.e., doing something that causes a harmful outcome) are viewed as worse than the corresponding omissions (i.e., doing nothing and the same outcome occurring). In fact, "a few subjects were even willing to accept greater harm in order to avoid action" (Spranca et al., 1991, p. 1).

Current military operations are conducted in a highly "accountogenic environment," in which decisions are generated not by a desire for the *right* outcome but, rather, by the desire that those in positions of power and influence over the decision-maker's future will *perceive* it as the right outcome (Eyre & Alison, 2010). In contrast to the first Gulf War (Operation Desert Storm), in which members of the media were kept at a distance (Alper, 2014), in Iraq and Afghanistan, members of the media are embedded within (and under the protection of) military units in the field. Embedded journalism (as well as social media) provides "real time" and "transparent imagery of life on the front line" (Kennedy, 2008, p. 285), but it also means that operational errors are much more salient and that members of the military are being held accountable in domestic courts for actions taken "in the fog of war." Military doctrine previously advanced the mantra "Do the right thing while no one is watching." This has now changed to "Do the right thing with the whole world watching." One of our interviewees highlighted the role of accountability in his own decision-making in that when presented with a need to take action, he froze, worried about the potential repercussions for taking the wrong action. Although driven by a clear moral concern, this aversion to error was also influenced by an overt cultural aversion to errors:

The culture at the time was, a fault intolerant culture, right, they were more worried about whether you followed the procedure right, um, than whether they were setting you up for paranoia. You're paranoid about the Afghans as much as you were about your boss coming down on you. . . . Um so yeah that was absolutely in my thought process, everything there was about you know trying to prevent bad things from happening, they were less worried about that it seemed to me than they were about accomplishing the mission.

Furthermore, drawing on what we know from literature on "choking," it is arguable that his cognitive resources were depleted from a preoccupation with accountability, meaning he was limited in the resources he did have to calculate this complex choice.

Uncertainty

As discussed in Chapter 2, how individuals make decisions in uncertain environments must be central to any valid theory of decision-making. Rather than just viewing uncertainty as a "sense of doubt," the SAFE-T model considers *where* this uncertainty specifically stems from. The SAFE-T model characterizes uncertainty as endogenous (about the event itself; Klein, 1993) and exogenous (stemming from the surrounding management and team processes; van den Heuvel et al., 2013), and it has a clear negative impact on decision-making. Endogenous sources of uncertainty include ambiguous information, time pressure, and risk (Orasanu & Connolly, 1993). Endogenous uncertainty prevents decision-makers from developing situation awareness (i.e., "What is going on?") and affects their ability to model prospective outcomes of courses of action (i.e., "What will happen if . . .?"; Klein, Snowden, & Pin, 2007).

In military operations, endogenous uncertainty is high, and the importance of navigating it is well known; the strategic (and tactical) operating environment is characterized by threats that are "diffuse and uncertain," and the decision-making environment is marked by its volatility, uncertainty, complexity, and ambiguity (VUCA; Franke, 2011). In more traditional terms, Clausewitz (1984) writes that "war is the realm of uncertainty; three quarters of the factors on which action in war is based are wrapped in a fog of greater or lesser uncertainty" (p. 102). Maj. Gen. R. Scales (2009) argues that today's conflict demands "officers who can lead indirectly and perform in an uncertain, ambiguous, complex, chaotic and inherently unpredictable environment" (p. 22). Similarly, the US Army Human Dimension Strategy (2015) emphasizes development of "soldiers and Army Civilians who are not just comfortable with ambiguity and chaos, but improve and thrive in even the most difficult conditions and achieve mission success" (p. 4). Although it is well known that soldiers must deal with omnipresent uncertainty, very little experimental research has investigated what happens when soldiers make decisions under conditions of uncertainty. Shattuck, Miller, and Kemmerer (2009) investigated the effects of differing levels of uncertainty in mission planning. They developed a mission planning exercise using real-word vignettes from veterans of Operation Enduring Freedom and Iraqi Freedom. They used nine vignettes in their study: Three contained ambiguous/missing information, three contained conflicting information, and three contained baseline (complete) information. Overall, they found that more uncertainty was associated with slower decisions in all conditions, meaning that those who took a long time to assess the situation also took a long time to decide upon a course of action (Shattuck et al., 2009). This implies that endogenous uncertainty in a military context has a similar effect on decision-making as it does in policing contexts: Uncertainty during the SA phase causes repeated and redundant requests for the same information, whereas uncertainty during the PF and PE phases leads to choice deferral whereby the decision-maker avoids implementing an action, adopting a "wait and see" mentality (van den Heuvel et al., 2012, p. 181).

Exogenous uncertainty has received even less attention from scholars of military decision-making. Exogenous uncertainty derives from confusion over one's own expectations or expectations of another's performance (van den Heuvel et al., 2013). Within team-based decision-making, poor role understanding reduces confidence (Shanteau, 1997) and self-efficacy (Bandura, 1997), both of which are important for goal setting and action planning (Olson, Roese, & Zanna, 1996). Poor role understanding can also erode interpersonal trust within a team, which can in turn affect confidence and perceptions of reliability regarding other team members' judgment and advice (Budescu & Rantilla, 2000). This concept is discussed in much greater detail in Chapter 7 on team learning. Endogenous uncertainty can also affect team cohesion, reducing team members' willingness to share (McKay, 1991) and seek information (Sniezek & Van Swol, 2001). Naturalistic research on police decision-making in a hostage negotiation scenario found that exogenous uncertainty was more common than endogenous uncertainty, and it affected the planning and execution phases of decision-making (van den Heuvel et al., 2013). Exogenous uncertainty is likely just as prevalent within military decision-making. Militaries often adapt during war, meaning that they revise their tactics, techniques, and procedures (Farrell, 2011). In Afghanistan and Iraq, there have been large-scale strategic and tactical adaptations of military forces (Catignani, 2012; Farrell, Osinga, & Russell, 2013; Kahl, 2007). Alongside this structural adaptation, there have been equivalent changes in the nature of operations that soldiers undertake. For example, soldiers deployed to Iraq as part of Iraqi Freedom and to Afghanistan as part of Operation Enduring Freedom found themselves undertaking extensive population-centric operations that they had not been engaged (or trained) in before (Nagl, 2012).

In addition, a majority of military operations are conducted in partnership with other multinational units with which individuals have varying degrees of previous experience. The North Atlantic Treaty Organization (NATO, 2005, p. 1) has recognized this issue, noting that the increasing use of ad hoc, multinational, joint military units has "brought forward issues such as critical team size, distribution of specialties, leadership, communication, cultural diversity and their impact on robustness, flexibility, and effectiveness." Ben-Shalom, Lehrer, and Ben-Ari's (2005) field research with members of the Iraqi Defense Force showed that many combat situations present the military equivalent of a "one-night stand" in that they "have a finite life span, form around a shared and relatively clear goal or purpose, and their success depends on a tight and coordinated coupling of activity." In addition, military operations in both Iraq and Afghanistan have largely relied on partnerships with indigenous forces. Bordin's (2011) interviews with members of the International Security Assistance Force (ISAF) demonstrate several sources of exogenous uncertainty. ISAF forces displayed low confidence in the ability of their Afghan counterparts ("The overall quality of the Afghan National Army cannot be intelligently described. It would benefit Afghanistan to disband the ANA and start over again" [p. 25]), resulting in low interpersonal trust ("While on patrol . . . I don't trust them" [p. 24] and "I wouldn't trust the ANA with anything, never mind my life" [p. 25]) and generating a perception of incompetence ("They are just about useless; genuinely stupid" [p. 25] and "We are interfering with Darwinian Theory!" [p. 25]). When considering more recent experiences in Afghanistan, it would also be viable to propose that the increased emergence of green-on-blue attacks (also referred to as "insider attacks"; Shortland & Hilland, 2013) would be a significant source of exogenous uncertainty in combat-related decisions.

Compounding these issues, even within the US Army, the makeup of squads and fireteams is constantly shifting. Platoon leaders may have only six men for a typical nine-man squad (including squad leader), and individual role players can get reassigned without notice to other units. This ever-evolving unit makeup can erode trust and confidence beyond the external forces described previously.

UTILITY OF THE SAFE-T MODEL

We have adopted the SAFE-T model as an analytical frame in which to place our own research on military decision-making. We do not seek to replace the MDMP as a formal doctrinal strategy, but we believe SAFE-T provides additional explanatory power by keeping focus on the decision-making environment, the team, and the process via which the team arrives at or, more crucially, does not arrive at a decision. Fortunately, the SAFE-T model shows a very high degree of overlap with the current

MDMP and another well-known military decision-making model, the OODA loop (observe, orient, decide, act). The OODA loop was developed by John Boyd (1996) and involves taking in observations of the situation (SA), making judgments of the situation and understanding what it means (developing situation awareness), and deciding (PF) and acting (PE). Thus, the SAFE-T model has high scientific validity because it reflects the phase-based process decision-makers usually adopt and is structurally similar to other proven models of military decision-making (Cohen, Freeman, & Wolf, 1996; Kaempf, Klein, Thordsen, & Wolf, 1996). Yet, what the SAFE-T model adds is a series of "error traps" that can be encountered along the decision path, and it explains how these errors traps can result in different types of derailed decision-making (van den Heuvel et al., 2012). Based on how the variables we just described affect decision-makers' ability to navigate uncertain, complex critical incidents, Figure 3.2 shows an expanded version of the MDMP model that integrates the derailment pathways highlighted by the SAFE-T model (and extended by Powers and Alison [2017]) and the affective, ambient, cognitive, and organizational factors that can increase the likelihood that decision inertia will emerge.

By integrating the military doctrinal process for making effective decisions at war with an NDM phase-based model of decision-making, we have arrived at a hypothetical model of military decision-making that details the ways in which military decisions can falter and the latent factors that can hinder an individual's (or group's) effective transition through the seven phases of the MDMP. The lens for military decision-making permits us to focus not only on the outcomes of decisions but also on the deeper view needed to unpack the dynamic, cognitive, social, and organizational processes that can lead to errors in outcomes (Alison et al., 2015). In the military, in which outcomes cannot reliably identify *errors* in the decision process, the SAFE-T model has more explanatory power by outlining the conditions that lead to multiple types of failure to act.

CONCLUSION

In Chapters 1–3, we have sought to provide the scientific "lay of the land" in terms of what we know about decisions—what makes them hard and what are the psychological consequences of experiencing decisional conflict. We have looked to doctrinal and theoretical perspectives on how people make decisions under uncertainty and in war, including the Army's MDMP and the widely applied RPD model of decision-making. Both of these have their strengths. The MDMP is well suited as doctrine because it forces deliberate planning for a mission, minimizing the biases and omissions that can occur when relying solely on intuition. RPD explains how soldiers can maximize their experience and use intuition effectively to identify a workable course of

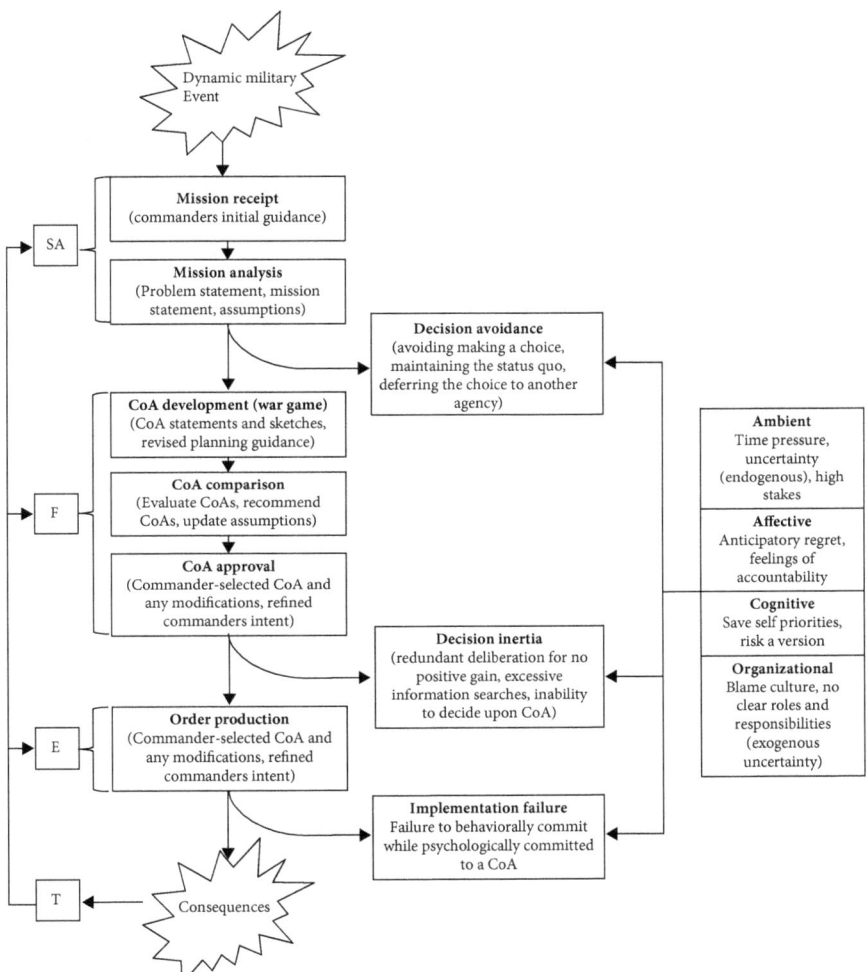

FIGURE 3.2 The integrated SAFE-T/MDMP model of military decision-making identifying derailment points and factors that can increase the emergence of decision inertia.
Source: Neil Shortland.

action. However, our research has encountered many cases of decision-making that cannot fully be explained by these models. From our studies, we know that military decision-makers are affected by organizational and social pressures, feelings of accountability, and anticipatory regret. Furthermore, they often do not have the luxury of choosing a "best" option and can often have very little equivalent experience (nullifying RPD approaches). These conditions are ripe for decision inertia to emerge. To explain these phenomena (and other kinds of failures to act), we have applied the SAFE-T model to understand how soldiers make complex least-worst decisions. We proposed this theoretical framework to help better understand situations in which critical time-sensitive decisions are stalled or avoided. We identified the pragmatic and scientific relevance of the model (in that it demonstrates clear overlap with the MDMP). Finally, and perhaps most important, we discussed the importance of decision inertia as the result of dynamic, cognitive, social, and organizational processes (Alison et al., 2015). We return to these concepts in later chapters. Military culture places a premium on timely, effective decision-making. In the military, inertia within the decision-making process can have significant negative consequences—even more so than in emergency responding.

In the next few chapters, we use the SAFE-T model as a framework to explore the psychological nuances that occur within each discrete phase of the least-worst decision-making process. Throughout each phase of analysis, we present real decisions made by soldiers along with the psychological underpinnings of each stage of the decision-making process. In doing so, we show how soldiers' decision-making can remain robust even in the face of intense derailment pressures that civilians might find crippling.

4

SITUATION AWARENESS

The unit enters upon the battlefield and moves across ground within range of the enemy's small arms. The enemy fires. The transition of that moment is wholly abnormal. He had expected to see action. He sees nothing. There is nothing to be seen. The fire comes out of nowhere. He knows that it is fire because the sounds are unmistakable. But that is all he knows for certain.... The men scatter as the fire breaks around. When they go to ground, most of them are lost to the sight of each other. Those who can still be seen are for the most part strangely silent. They are shocked by the mystery of their situation. Here is surprise of a kind which no one had taught them to guard against.... In essence, it is against this very situation that his unit must find the means to rally if it is to succeed in battle.
—Col. S. L. A. Marshall (1947, p. 47)

Col. Marshall's quote evokes in us what it means to be lacking *situation awareness* in combat. The first phase of the SAFE-T model (and many models of decision-making) is gaining an understanding of what is happening. In this chapter, we explore the process of sense-making to gain situation awareness in combat. We present cases from those involved in offensive operations in the field, as well as those involved in more remote operations whose sense-making is solely reliant upon small snippets of the scene that they can observe through the technology available to them. We discuss the limitations (cognitive and situational) in trying to gain an accurate picture of what is happening on the ground and how this complicates the remaining phases of the decision-making process. We pay special attention to the role of cultural differences, the difficulties in making sense and "storytelling" in environments that have little in common with one's own. Finally, with reference to a real case of mission planning in Afghanistan, we examine the delicate interplay of situational awareness and action.

SITUATION AWARENESS

Most decision-making models across the psychological literature (including the SAFE-T model) involve the same basic steps: Build situation awareness by understanding what has happened, why it is happening, and what will happen in the future. Based on this prediction, identify or generate a series of options that we could take to achieve a given goal (Fellows, 2004; van den Heuvel, Alison, & Crego, 2012). Thus, perceiving the situation is the preliminary stage of any decision. However, what happens when we cannot gain situation awareness or when two opposing, equally likely, "stories" can explain the situation in which a soldier finds him- or herself? In such situations, a decision-maker is often presented with a least-worst decision, not because of the nature of the environment itself but, rather, because the decision-maker's sense-making is so impoverished (due to high uncertainty) that he or she is unable to know the outcomes of his or her options. It is clear here that storytelling will not get a soldier closer to the truth of the situation.

When decisions are equally adverse because of impoverished situation awareness, soldiers continually seek more information, even though neither the time nor the information is available to remedy this uncertainty. People are driven to try to reduce ambiguity, even at the expense of competing goals. It is important to stress this point because we are not saying that additional information should *not* be sought, but we are saying that in most time-sensitive situations, there comes a point at which the decision-maker has to accept the (uncertain) situation awareness he or she has and *commit* to a course of action. Decision inertia thus mirrors the economic concept of "diminishing returns," in which (Shepard & Fare, 1974)

> [when] equal quantities of capital and labor are applied successively to a given plot of land, the output resulting from these applications will increase monotonically at first up to a certain point, after which further applications will result in steadily decreasing product increments tending to zero. (p. 287)

Similarly, there is a point at which additional searches for information no longer provide any "gain" in terms of improved situation awareness. Instead, (redundant) searches for information negate other gains such as the "initiative," "element of surprise," and "operational tempo." This is vital because in war, *when* something is done can be as important as *what* is done.

SITUATION AWARENESS AND SENSE-MAKING

Situation awareness (SA) most simply is "knowing what is going on around you" (Endsley, 2000, p. 2), and "sense-making" is a central part of this process. Sense-making is a motivated effort to understand connections in order to anticipate future trajectories and act effectively (Klein, Moon, & Hoffman, 2006). Sense-making involves sifting through large volumes of information and selecting the necessary pieces (Ben-Shalom, Klar, & Benbenisty, 2012). Humans are guided by the cognitive edict to make sense of what is happening in their immediate environment. Sense-making is one of the fundamental roles of perception and cognition, and when it fails, the motivation to act can unravel, leaving the individual in a state of "limbo." This need to "make sense" of our surroundings (and the negative effects of not achieving this) cannot be underestimated. Chater and Loewenstein (2016) argue that sense-making is a powerful human motive, positing "the existence of a 'drive for sense-making' which, we argue, is analogous to better known drives such as hunger, thirst and sex" (p. 137). Innate drives are self- and species-preserving. They are biologically determined in that they impose "on every civilization and on all individuals in it the carrying out of such bodily functions as breathing, sleep, rest, nutrition, excretion, and reproduction" (Malinowski, 1944, p. 7.51). Innate drives are usually accompanied by powerful positive or negative affective states that cause individuals to approach or avoid stimuli and situations; for example, being full is pleasant, whereas going hungry is not. These affective states provide the motivation to act. We do not need psychologists to tell us that uncertainty, the outcome of unsuccessful sense-making, is unpleasant (Shaw & Thomas, 2013).

There are many definitions of SA, most of which are closely linked to aviation, but a generalist definition couches SA as "the perception of the elements in the environment within a volume of time and space, the comprehension of their meaning and the projection of their status in the near future" (Endsley, 1955a, p. 36). Endsley (1995a) proposed a descriptive model of the SA phenomenon. Her theory of SA sought to "explain dynamic goal selection, attention to appropriate critical cues, expectancies regarding future states of the situation, and the tie between situation awareness and typical actions" (p. 34). Endsley (1993) proposed that the first step of achieving SA is to perceive "the status, attributes, and dynamics of relevant elements in the environment" (p. 36). This entails getting the best information possible on the relevant attributes of the environment. For example, a golfer would want to know his or her distance from the pin, the wind conditions (and if this will change), and the shape and slope of the green. A tactical commander, on the other hand, needs accurate data on the location, type, number, capabilities, and dynamics of any potential enemy forces in a given area. The second stage of developing SA is to comprehend the situation. Comprehending the situation is based on synthesizing the elements collected in Level 1. As such, in line with Gestalt principles, Level 2 SA "goes beyond simply being aware of the elements that are present to include an understanding of the significance of those elements in light of pertinent operator goals" (Endsley, 1995a, p. 37). To unpack this, the decision-maker forms a holistic picture of the environment. This is the sense-making component of SA. The final level of SA is the "ability to project the future actions of the elements in the environment—at least in the very near term"

(Endsley, 1993, p. 37). What this level of SA entails is therefore being able to predict the outcomes of a choice given the known causes of current elements within the situation. Endsley's model is the most extensive and highly cited model of SA (Golightly, Wilson, Lowe, & Sharples, 2010), and it has immediate applicability when examining the way in which members of the Armed Forces are continually required to encode the elements of a situation, understand their interrelation, and use this to predict whether they need to take (potentially life-saving) action.

Here, we present the case of one soldier who we interviewed, in which all aspects of Endsley's SA model are demonstrated. The reader will, we hope, be able to see how the soldier, in real time, works through the process of identifying what is going on, attempting to understand what is causing it, and projecting what could happen in the future. In this scenario, we hope the reader can appreciate our previous point that a lack of SA alone is often sufficient to create a least-worst option, when, in reality (and with perfect information), there is one correct course of action.

THE WALK-IN

In 2011, Maj. R. Webb was at dinner one evening when he received a phone call from a number he did not recognize. He answered the call, and a man with an unidentifiable accent (although he later identified this as a Pakistani accent) said he had "intelligence he wanted to give to the Americans." Maj. Webb was not familiar with "walk-ins" (when a member of the public voluntarily offers information); his role was more of a "fixer" on the base, doing whatever his boss needed him to do—mainly organizing meetings and solving logistical issues. As Maj. Webb phrased it,

> If there was something to do I would try to figure out a way to get it done. And that served me well during my first couple of years when I was deployed. When my boss, my boss was a colonel working for the four-star Afghan General, if he needed anything, didn't matter if I knew how to do it, I was going to find a way to do it, because you know I had a four-star general on the line. Everything I did on that whole deployment was new to me, it was all novel things that I had to figure out on the fly. And I was happy to do it. So, I had become used to trying to do things that were outside my comfort zone and I think that probably came into play as well, is the facility with which I was willing to accept doing unusual things that I wasn't accustomed to doing.

When he received this call, Maj. Webb was with a Naval intelligence officer, and after discussion they decided to set up a time and a place to meet with the walk-in. So, the two of them started talking with security forces and intelligence officers trying to find individuals who were both able and willing to help them go outside the wire to meet and collect the walk-in. Unfortunately, when they identified the specialist unit associated with such situations, they were informed that although the unit could conduct the interrogation, it did not have the ability to do the walk-in (i.e., getting him off the street). The unit did not have body armor or helmets. So, Maj. Webb designed a basic plan in which they were going to meet him on the street and he was going to park his car in a very specific place. The walk-in was then supposed to give a preapproved signal, raise his hands, and allow them to do a routine search before bringing him into base. Maj. Webb and his Naval intelligence colleague set out to meet the walk-in.

They arrived at the meeting site, and Maj. Webb took position providing security for the left side of the street. However, rather than follow the agreed upon course of action, the walk-in simply walked up to Maj. Webb and announced, "Hey, I've got information for you." So, they reorganized their defense and put the walk-in in a "position of disadvantage" (on the ground, with his arms spread out) and started patting him down. Across the street was a small SUV. Maj. Webb recalled that they "didn't think anything of it," except that it was "unusually nice" for Afghanistan, where "almost everything is dirty." The SUV also had dark tinted windows, which, again, was unusual in Afghanistan. During the pat-down, however, the lights of the SUV suddenly came on, and the SUV moved across the street directly toward them.

At this point, we would like you to assess the situation. Level 1 SA has been provided for you: a walk-in on the ground and an unusual-looking SUV driving toward you. What are the possible explanations for the behavior of the SUV and, crucially, what is going to happen next? Now, when you consider possible different explanations, you will see that potential courses of action are quite different, and the cost of an error is quite high. This really is quite a classic least-worst scenario when it comes to Afghanistan. At the simplest level, there is one question: What do you determine the intent of the driver to be? If hostile, the best thing for you to do is take offensive action. If you deem the driver's intent to be benign, then the best course of action is to remain calm. As Maj. Webb recalled,

> Now at this point, and you talked earlier about a point at which there was no good decision, I don't know what their intent is, I know they are coming straight at my guys, at my friend, um, and I don't know what I am supposed to do right? I don't know if this guy is a danger or not. And I also don't know, with a SUV you've got a fair amount of cargo capacity there, so I'm thinking this could also be a bomb? So inside my head, I am thinking "he's going to die," this guy is going to blow himself up, because, just the way he came out,

he really tore his way out of that parking lot and came right at them . . . and there was a split second when I thought I can save [the walk-in's] life right now by possibly murdering an innocent person. And on the other side of it, and I had to weigh that against, do I want my friend to possibly be killed, and at the same time I'm thinking, it's possible that if I shoot at this guy, it isn't going to save my friend anyway right?

So how did Maj. Webb assess this situation, and to what degree was the process through which he assessed the situation reflective of Endsley's model? First and foremost, he gained all of the relevant information. However, it is interesting to note that not all of the situation was attended to *equally* during this first parse. For example, Maj. Webb did not really focus on the SUV because he was preoccupied with the mission, identifying and securing the walk-in and providing rear security for the team. However, when the SUV surged toward them, Maj. Webb reanalyzed the SUV, identifying new and relevant details such as the blacked-out windows and the general level of cleanliness (compared to what he was used to seeing during his time in Afghanistan). This reanalysis corresponded to Endsley's second stage—comprehension—in which the significance of these elements is understood in light of the operator's goals (in this case, protecting his men, the walk-in, and himself). In this case, then, Maj. Webb took elements of the situation (the tinted windows and the SUV) and used them to generate an explanation for the events that were ongoing:

He was so out of place, the vehicle was out of place, his tinted windows were out of place, his speed, the aggressive nature of his motion was out of place, um, everything that happened at that point gave me a bad feeling, um, of imminent danger and I had spoken with other people, um, and I had friends who were in a vehicle that was hit by a human-borne IED [improvised explosive device], so basically someone walked up from their periphery and walked right into the vehicle and blew themselves up. They were fine, they medevac'd him but they were fine, just shrapnel injuries and things like that, nothing terrible. The guy of course was killed. But the one thing they said to me about that, because I talked with them about the incident, but the one thing they said about it was that it looked out of place, he looked out of place, he was wrong in that environment and they said that, you know, if you've been there no one's white, no one's white, everyone's wearing white clothes but no one is white, they are always dirty and this guy was in pristine clean white, he was, you know, mumbling to himself, probably praying, as he was walking and he was going straight at the vehicle without looking at either side. So, they said everything about him looked out of place but it happened so fast they couldn't react, and so that was in my mind you know through most of my time there was find the thing that was out of place, and that was out of place, that vehicle was out of place.

Now, having sat with Maj. Webb and discussed this event for several hours, we are fairly confident that his assessment of the situation leant toward it being threatening and there being a need for offensive action, but other things prevented him from taking offensive action. As he stated during our interview,

That was all uncertainty on my part, I felt like that might be a threat, but I wasn't sure enough if it was, right. On the one side, you know I've got a deadly weapon and intent, I just don't have the know-how, I just don't know whether I should be firing or not. And so, I guess some part of me is worried about killing him, and about escalating a situation that was non-hostile. And so, that was probably a larger factor in my mind and, um, I'm thinking I don't want to screw this up, and so I'm really going to pay attention to what they are doing, and um, I think if they had been, you know, he was drawn on the guy and I was drawn on the guy he was ready to fire and I was ready to fire. If, um, we had heard, if a round had gone off I would have been firing immediately, no question about it, if you know the driver had shot or my guys had shot I would have fired without hesitation, but I didn't know, I wasn't sure if I was supposed to fire.

In terms of Endsley's model, then, Maj. Webb achieved Level 1 SA—he perceived the elements in the environment—but his issue was Level 2 SA, comprehension, because he did not know the significance of these cues in terms of his goals (and from this he could not establish Level 3 SA: predicting future outcomes). A central reason for this is that although he could detect salient cues, and even use these to diagnose potential intentions of the SUV driver, he lacked the *confidence* to determine their significance. Furthermore, this confidence threshold was high because of the accountability and low mutability of the choice (as predicted by the SAFE-T model). Hence, he knew the significance of what he was seeing, he just did not know if the elements were significant *enough*. This issue emerges frequently in our discussions with soldiers. Although often we view that insufficient or too much information causes uncertainty, sometimes the more pressing issue is diagnosing what this information means. As Maj. Webb stated, "I was trained to look for things that were unusual to me, and that seemed unusual to me, but I wasn't sure if it was unusual enough."

USUAL *ENOUGH*

One factor that hindered the SA of Maj. Webb was that he had very little experience of walk-ins. As he noted,

> I feel like if I had seen more of these walk-ins I would have been able to know what was unusual and what wasn't. How unusual was it and how wary should I be based on how this went down? Being that this was my first and only one, I don't know. I don't know how experience would have changed that. I suspect that I would have felt more clear about what I was to do at the time.

This is not uncommon. As David's (1997) study of military errors found, a "typical handicap seems to be a lack of command experience. Naivety tends to promote vacillation and over caution, resulting in lost opportunities and ultimately disaster" (p. 1). Experience aids SA. This is one of the most commonly found results in SA research (Klein, Calderwood, & Clinton-Cirocco, 2010), and it has repeatedly been shown that experts have different perceptual and cognitive mechanisms compared to novices, and these facilitate their decision-making (Klein & Hoffman, 1993; Randel, Pugh, & Reed, 1996). Researchers in many areas (especially naturalistic decision-making) have found that expert decision-makers can generate enhanced SA in that they can clarify and understand a situation (Klein, 1993; Klein, Calderwood, & Clinton-Cirocco, 1986)—mapping on to Level 2 situation awareness; they comprehend meaning. In fact, one could easily recast expertise as precisely the ability to pattern match to develop better Level 2 SA. As Klein (1989) found, experts, rather than novices, spend the majority of time on a new problem in acquiring the information necessary to perceive the elements in the environment (Level 1 SA) in order to understand them (Level 2 SA), then accurately predict what will happen next (Level 3 SA) to identify the appropriate solution from memory. Time and time again, our interviews reinforced the importance of experience in how the soldiers perceived an ambiguous situation. One interviewee recalled,

> Yeah. Experience. I've noticed that through the different deployments, the decision making process gets easier. The decisions don't get easier, but the process of elimination, like I said exercising patience, your world gets a little bigger, you understand that there are more things you can bring to bear on the problem, um, so that process, and that visualization, I, um, I mean, on the first deployment, you're still doing the visualization, but it's wilder. You don't know exactly, you think of the worst thing that can happen, but sometimes you have no idea the worst thing that can happen, because there are a lot of

terrible things that happen. So, I would say experience, and then we trained pretty hard going in. I mean, we put our commanders. We put hard, unsolvable problems in front of them that kind of caused everyone to exercise this, and then we usually throw in ethical dilemmas on top of that, and cause us to go through that. Um, so I think the training and the experience certainty helps. And I've noticed that through the years, different positions, different deployments, doing different things, the decisions actually get harder as you go up in rank, but the process to make the decisions, I seem to be more comfortable with that. So, it was training and experience that played a big role.

Expertise profoundly affects an operator's SA, increasing the ability to attend to relevant cues in the environment, see patterns, and make sense of the situation, as well as better project into the future what is most likely to occur, given the stimuli available to the decision-maker at that time. In short, expertise improves SA at each of Levels 1, 2, and 3 by improving the operator's ability to rapidly assess what is happening in his or her environment.

Thus, the foregoing clearly shows that experience in the field leads to better decision-making, as predicted by the recognition-primed decision-making approach. The issue as we see it, however, is that there are several barriers that prevent members of the military from developing expertise in the same way that firefighters or police officers do. Soldiers, across their career, will operate across a series of diverse operational theaters, and indicators of threat in one theater may not match (or worse even, may contradict) those in another theater. Instead, they must rapidly develop expertise the minute they land in theater. Furthermore, despite incredible efforts to train them, soldiers often express that training can never quite capture the reality of war. This point is important to us because many of our conversations with soldiers have highlighted a gap between training and reality. To put this point in perspective, we next outline the case of a drone pilot who was required to make a similar decision as Maj. Webb had to make. This narrative is especially important here because of the differences he highlights between the decisions he trained for and those he was required to make on deployment.

"IT DIDN'T SAY IED ON IT"

Drone pilots provide the "bird's-eye" view (Level 1) SA for soldiers on the ground. From their position, they are able to spot insurgents laying IEDs as well as identify insurgent positions to aid on-the-ground targeting. They can also engage targets themselves. Although drones are referred to as "unmanned," this is a misnomer because there is a human decision-maker in control of each drone.

In military terms, there is still a "human in the loop." Drone pilots must ensure they establish good SA of the battlefield, despite their physical distance from the target.

In 2013, while working within the eastern part of Afghanistan, Airman First Class R. Evans (AFC Evans) received orders from his coordinator to conduct a counter-IED job—basically "go to this location and see what these people are doing there." So, he was given the grid location and told that there were reports "that people were seen burying stuff over by a main highway through Afghanistan." Originally, AFC Evans thought this would be pretty simple: "They would immediately know it when they saw it," and when they got to the area, it did not take them long to see people digging and burying. They knew people were burying *something* because of temperature differences between the ground and the regular soil, but they could not get a clear picture as to what it was. As AFC Evans stated, "A camera is only as good as it can be, from thousands of feet up in the air and through the haze and dirt of Afghanistan. You can only see so much" and the "surface of Afghanistan is like the surface of Mars." So, suddenly this simple decision was not as easy as in training where you "see someone burying something and then running a cord across the road." Even when they saw a motorcycle and a car drop something off, "you couldn't see what it was. It didn't say IED on it. It was a very blurry picture." If they were burying IEDs, AFC Evans was allowed to use "deadly force" against them (meaning he could launch a missile strike at their location). They did not have enough information to determine what was going on—to know exactly what the people were burying and, specifically, whether or not it was an IED. In AFC Evan's words, "So what choices do I have? One choice of firing on people who, potentially, were planting IEDs in the ground, you know 'bad guys' or another choice, fire on innocent people doing something random."

The latter was the greater concern for AFC Evans; Afghan civilians, in his mind, "did all sorts of whacky stuff. . . . I don't know if you have seen footage of day-to-day living in Afghanistan on the road, but they do some pretty whacky stuff on the highway, or riding horses down the highway." AFC Evans recalled that his first mission in Afghanistan was to look for anything suspicious: "Well everything looks suspicious, that's just how they are." He stated,

> Yeah just the fact that we had not seen this type of activity before like I personally had not seen someone burying something you know? Whatever it was, I had not seen that so obviously that would get your attention and would raise the alarms because it's not a day-to-day activity. It's not like pushing barrels down the roads and various other things we had seen so just because

of that it pushed me [toward thinking they were laying a IED] but again we couldn't give a '100% of that's what it was.'"

The issue was that AFC Evans had very little points of reference from which he could use to determine what these bits of information (Level 1 SA) meant (Level 2 SA) and what, possibly, could happen next (Level 3 SA). AFC Evans stated,

> When I first got there we were inexperienced and even those going to Iraq and stuff had not been to Afghanistan so we were told to look for suspicious activity and it's a lot of stress 'cos it's like that looked suspicious but it took time to flesh out the norms for that area versus what's really terrorist activity so this had taken place later on when we were almost done in there, so that played a lot. If this had been when we first got there it might have pulled a decision one way that it didn't later on . . . and I know that contradicts what I said before 'cos I know I said I'd never seen this before but the baseline: People do things completely different to us so don't jump onto everything you see. It's a different world so. . . . It flushes a lot of things out that you would jump on and threat detection and what not, you flush a lot out so you focus on threat detection on hills, and even activity that happened on hills is sometimes normal 'cos you have goat herders and hills and these are like . . . from when we first got there trying to jump on everything you see, they still don't look normal to you, but I'll never say that it's normal to ride horses down the main highway.

And, in many cases, the level of realism afforded in training is simply insufficient to reflect what a "threat" looks like on the ground. As one of our other interviewees recalled,

> They train you that, you know, for example, you're doing convoy ops and you get hit with an IED. They will literally take a 155 artillery round and they will toss it in the middle of the road with wires hanging out of it, and it is never like that. Never like that. You're never going to see it, you know, um, I would say that they train you that everything is a threat, but realistically, I remember my first couple of times going out, "there's black flags everywhere what are those," you know, "are they signaling something?" Pigeons, a lot of people had homing pigeons, you know pigeons getting up? Oh, are they signaling the convoy or whatever? And sometimes I think it was, but a lot of times I think it is just pigeons flying around—or they are just flags! So you know, they train you that the threat. . . . You know, every single time we train, you always get hit, or someone gets killed, or someone gets shot, you got to react in a different way. So you know, training perceives it as more often there

than it is. I think obviously the threat is there, and you really got to kind of learn it.

After much deliberation, AFC Evans decided not to fire, averting a serious and very costly error. After conversing with the Afghan National Army, they found out that (1) the individuals were burying fruit for a local farmers' market and (2) burying fruit overnight to keep it fresh was commonplace. What we see here, then, is the important role that understanding cultural practices and norms plays in SA and specifically the additional challenge this adds for military decision-makers compared to their police and emergency service counterparts.

SA GONE AWRY: ANATOMIES OF AN SA ERROR

Developing SA is a fragile process. As demonstrated previously, good SA requires an understanding of local (and cultural) norms, the ability to attend to the most relevant information, the ability to extract meaning and interrelations from the many facets of a scene, and the ability to project into the future to determine what will happen (and, more crucially, what will happen if you do "X" or "Y"). Given this, it is not unsurprising that in uncertain, high-risk environments there are often errors in SA. Jones and Endsley (1996) proposed a three-level taxonomy for classifying and describing errors in SA. The first level of failure is to "incorrectly perceive the situation." In these cases, the information may be available to the decision-maker but is unattended because of too much irrelevant available information. Issues such as workload can further confound this by decreasing working memory capacity and, hence, decreasing the number of "slots" available in consciousness to process the relevant information (Hockey, 1997). In other cases, information may simply be unavailable or too ambiguous. Orasanu, Martin, and Davison, (1998), for example, in their study of 37 aviation incidents in which crew behavior caused an error (as determined by the National Transportation Safety Board), discovered that the most common error was to "continue with the original plan of action in the face of cues that suggested changing the course of action" (p. 3). Furthermore, these plan-continuation errors accounted for approximately 75% of all tactical decision-making errors. This means that the majority of errors stemmed from an inability to either (1) update one's SA or (2) re-evaluate an already selected course of action based on a new SA. This astounding statistic underscores the difficulty inherent in quickly assimilating unfolding information and using that new knowledge to change one's behavior, and this literally caused most of the aviation crashes observed.

There are several cases in which, under such conditions of uncertainty, an error in judgment has been made based on incomplete Level 1 SA. During Level 1 SA, we can misperceive the situation due to the influence of prior expectations (Jones & Endsley, 1996). Prior expectations influence all of us all the time. For instance, Wansink, Payne, and North (2007) found that labeling the same wine as either from California's Napa Valley or from North Dakota had an influence on (1) how much people *expected* to enjoy the wine, (2) how much they *did enjoy* the wine, and (3) how highly they rated the flavor of cheese that was served with the wine. Given identical sensory input (in this case, the same wine), people allowed their prior expectations to modify their actual perceptual experience. The following case shows how easily we can see what we think we should see.

"HOW DO YOU KNOW THEY ARE MILITARY? THEY COULD BE WOODCUTTERS?"

Col. Barker served with the US Army for 25 years, and in our interview he spoke about what he called "the most incredible decision he had ever made." In early 2009, Col. Barker was a squadron commander stationed at a combat outpost in a very mountainous area in southern Afghanistan. This place had been continually rocketed by the Haqqani network (one of Afghanistan's most experienced and sophisticated insurgent organizations). It had been rocketed almost daily, and although they knew the point of origin of the attacks, they had been having a lot of trouble tracking down the command and control and logistics bases. Furthermore, because it was so mountainous, they could not get troops up there to do a thorough search and attack. At one point, a rocket even hit the fuel farm inside the base, at which point Col. Barker thought that the whole base and all his men had been engulfed in flames. At this point, they were "getting pretty desperate":

> Most of it was cat and mouse kind of thing, a lot of mountainous terrain, too much to put someone out to observe. So it was almost the same time every day or every other day, probably late morning and 9, 10ish rockets start firing at the American base. So after several days of this in the morning we would start putting you know UAVs [unmanned Aerial Vehicles] over the area, we also had like a counter mortar rocket radar that we would put in the area ... and you know when the rockets went off we would have to do our best to find the point of origin and use counter rocket radar and then fire artillery at that base to try to neutralize it. And this was almost a daily occurrence, rarely did 2 days go by when we didn't have rocket attacks on the wilderness.

At approximately 2100 one night, Col. Barker was sitting in his office when some Special Forces operators came in and said that they knew where the logistics base was located. The source was a brother of an insurgent; they had him on the phone, and he said he could locate it. During the next 45 minutes, they talked to the intelligence source and the Special Forces troops for a while and eventually narrowed the location down to one spot on a mountain. A critical issue was the credibility of the source, so they sought to understand who he was and his motivation for giving up this information: "You know. He's got a brother up there. Why do you think he is informing on his brother?" A further hour was then spend developing the target, talking to the source on the cell phone, using the UAV to interrogate the target. A lot of this was Col. Barker "looking at a screen, looking at something fuzzy. Trying to frame it into something that I know I can fit into; you know, that's not a camp around there, that's military activity and a logistics base and that sort of thing":

> So you know the field of view on this thing, and especially at night is like looking through a soda straw, you never get a good picture. You get a decent picture when it's a narrow field of vision but when they pan out the resolution just you know it goes to crap and you can't see anything so. So basically what I'm seeing is a wooded area, so it's a hilltop, it's very rugged and there's an uneven plateau up there. It's wooded, extremely wooded, especially for Afghanistan and they are really up in the high mountains so this was maybe 11,000 feet of elevation. So very very high. It's kind of an uneven plateau. Wooded, so you're having a lot of interference from the foliage, but you can tell what it looks like. There's a tent but you're only seeing it because you're not seeing anything around it, it's kind of a fuzzy shape and you can see guys walking in and out of it, walking away from it, using it as a central point. There were no camp fires that I could tell easily and those are usually a pretty distinctive signatures. And most of the time you know if you see a forest fire the guys aren't trying to conceal it so that's what it looks like. Like there's a tent or something in there and guys are using it as a central point moving back to other locations. So that's what I saw.

Although it was night, they thought they had located the site, in and around the area they were told it would be. Furthermore, this activity seemed to be in a location that "no legitimate activity would take place in." So, Col. Barker had two streams of information: a source (who the Special Forces believed was credible) and he could look at it himself and see what he thought was activity. Given this, he thought he had a valid military target that he could attack "and that was the decision point when I decided to launch the mission." They had good SA, which increasingly supported Col. Barker's assessments:

> You can see guys moving back and forth that the I guess you wouldn't expect to see if guys were camping out. You don't expect guys to camp out unless

they have a good reason in Afghanistan, outdoor recreation is not what they do. But it was just the level of activity and to me it looked purposeful and those guys are going back and forth like there was some kind of military operation going on in that area so that's kind of what convinced me to do that, plus the tent. I thought you know okay well it's kind of what we would do, set up a command post and have a lot of stuff going back and forth especially if it's a logistics base.

However, there were also a few missing bits of data that, in an ideal world, would have been available but whose absence was easily explained away:

One thing that I tried to do that I couldn't do was I tried to listen to their frequencies and tried to do a little bit of electronic warfare. Right before a rocket attack you would get a spike in hand held communications and then you knew "right okay something's gonna happen, there is an IED going off or a rocket" but that's the one thing we couldn't pick up that we sort of would have expected. But we discounted that; they're all on one base, they won't need their radios, they are not doing operations, they are down for the night or preparing for operations the next day.

Thus, Col. Barker started planning the attack. He had US Air Force bombers available to him and artillery from his own base. He spent a couple hours putting the plan together, and they then conducted the attack on the base, hitting it with bombs and artillery. They are watching it at the same time, and "the minute the bombs start hitting, it looked like you kicked over an ant hill, stuff went well everywhere." There were a lot more people at the camp than they had estimated. They continued to attack it:

Know the attack probably lasted 30 minutes you know so erm . . . and the mission so 30 minutes we're hitting it with bombs and artillery and looking at seeing stuff kind of move around and then refining targets and attacking it again with artillery so probably about 30 minutes, that may be a little bit generous actually.

Approximately 10 minutes after the end of the mission, Col. Barker got a call from his boss, the Brigade Commander, asking "What are you doing?" Col. Barker replies that he is attacking a military target. The Commander replies, "How do you know they are military? They could be woodcutters?"

The realization came into my head that there was possibly an alternative explanation for why there are a bunch of men on a mountain top in Afghanistan. So you can imagine the sinking feeling where I think instead of righteously killing the insurgents I may have just mass murdered a bunch of innocent civilians. . . . We talk about it, and obviously he is upset, and I just hope I'm not a murderer.

At approximately 0600 the next day, Col. Barker sends one of his companies up into the mountains to conduct a battle damage assessment. The mountainous terrain made the search very difficult; however, after being sent back up the mountain (after the first search found nothing), at 1400 the platoon discovered the site, within which there was a lot of ammunition and communication equipment. It was "obviously an insurgent camp," which was a huge relief for Col. Barker.

Although this decision resulted in a positive outcome (the removal of an insurgent camp that was launching a significant number of attacks against the outpost), Col. Barker's decision-making still warranted further examination. As discussed in Chapter 1, there is a difference between good decision-making and good outcomes. Fortunately, in this case there was a good outcome despite poor cognitive strategies being employed in the decision-making phase. Even Col. Barker agreed that his decision-making was, from a process standpoint, less than ideal:

> What I didn't know is that because of my decision, my boss and his boss were talking that night about relieving me for just for doing that without having a real good reason to do it. Looking back on it over a number of years and even doing some leadership and decision making courses I don't, I can never understand why I made that choice. . . . I saw what I wanted to see you know. I think I remember until the day I die the sinking feeling I had when my boss said "Why don't you think they are woodcutters" and that's a situation we had had before in those mountains because a lot of illegal woodcutting had gone on there and you know I don't think it ever came into my mind that they could be woodcutters and I could just be wrong about what I was seeing so you know I was highly disappointed in myself for not thinking about you know alternative scenarios that could have fit the evidence.

Confirmation bias, introduced to psychology by the Nobel Prize-winner Daniel Kahneman and colleague Amos Tversky, refers to the idea that once we have hit upon a single explanation for something, we search only for evidence that supports it. The psychologists Hugo Mercier and Dan Sperber (2011) took this one step further when they suggested that reasoning was a mechanism designed by evolution not for arriving at the truth but, rather, for systematically finding arguments that justify one's beliefs or actions. Col. Barker noted the role of confirmation bias in his decision-making process:

> Quite frankly, it is what I wanted to [see] once I started looking at the pictures and correlating what the source was telling me you know it's like okay that's

what I want to see, it was some confirmation bias going on or okay yeah that's ... I see what I wanna see and I think I discounted other explanations of what it could have been you know.

Thus, the process of interpreting the situation correctly is highly complex, stemming not only from exogenous variables (e.g., missing information or conflicting intelligence reports) but also from endogenous variables such as tiredness, stress, emotion, and the weight of prior expectations. As Col. Barker highlighted, he was tired and frustrated after a series of efforts to prevent a series of increasingly lethal insurgent attacks had failed. This can directly affect the way in which situations are perceived, resulting in incorrect courses of action and potential decision errors, the results of which can have long-standing implications for the soldier (the effects of making such decisions on the psychological well-being of the soldier are discussed in Chapter 9).

Each type of SA error (misperception, a lack of comprehension, an inability to forecast, and an inability to update SA based on new information) alone can have devastating consequences for decision-making. But SA errors often interact, causing decision errors. A notorious example of SA errors in war is the situation assessment that led to a drone strike on February 21, 2010, in Afghanistan that resulted in the death of 15 civilians (and wounded 12 more). Usually, studying decision-making in the field relies on post-event recall; however, following a *Los Angeles Times* investigation, the transcript of the conversation between the drone pilots involved is now available, giving us insight into how SA is developed in real time. Despite knowing the outcome of the drone strike, we ask the reader to join the soldiers in trying to establish Level 1, 2, and 3 SA throughout this case and as more information unfolds. In doing so, we hope you will be able to understand how, with high enough uncertainty, each element of Level 1 SA can be used to create alternative understandings of the scene and, hence, completely different interpretations and plans of action.

"IS THAT AN *EXPLETIVE* RIFLE?"

Just after midnight on February 21, 2010, a convoy of vehicles was heading down a mountain road in central Afghanistan. The convoy was watched by US military surveillance, via infrared camera, by a Predator drone crew in Nevada, who screened footage to Florida, an airplane crew, and an American special operations unit on the ground nearby. For more than 4 hours, these units tracked the convoy, trying to decide whether it was friend or foe. During the early stages of the operation, attention was focused on three vehicles 3 or 4 miles from the special operations unit on the ground. Several factors raised suspicion: The operators cite that two of the vehicles are "flashing lights" and "signaling" between each other. In addition, the Joint Terminal Attack Controller (JTAC) said there was a "demonstration of hostile intent tactical maneuvering," which in conjunction with the intercepted communications "would appear that they are maneuvering on our location and setting themselves up for an attack." Although they could not provide positive identification (PID) that the convoy contained weapons (they could see "possible mortars"), at this time the ground commander's intent was "to destroy the vehicles and the personnel." The vehicles pulled up to a local compound, and the operators agreed they could see approximately 12 personnel and some "lookouts." However, they could still not PID any weapons:

00:45 (Pilot): Go back to that guy down there.
00:45 (MC): See if you can zoom in on that guy, 'cause he's kind of like . . .
00:45 (Pilot): What did he just leave there?
00:45 (Pilot): Is that an *expletive* rifle?
00:45 (Sensor): Maybe just a warm spot from where he was sitting; can't really tell right now, but it does look like an object.
00:45 (Pilot): I was hoping I could make a rifle out, never mind.
00:45 (Sensor): The only way I've ever been able to see a rifle is if they move them around, when they're holding them, with muzzle flashes out or slinging them across their shoulders.

Intelligence from higher command was supporting the view that there were insurgents in the convoy, stating, "What we're looking at is a QRF [quick reaction force]; we believe we may have a high-level Taliban Commander." With this information in hand, the pilots reanalyzed the scene, making estimations as to which person they thought was the senior Taliban Commander:

00:55 (Pilot): Wouldn't surprise me if this was one of their important guys, just watching from a distance, you know what I mean?
00:55 (Sensor): Yea he's got his security detail.

Despite this, however, they could still not get PID on weapons and were relying on "tactical movements" to determine intent. In addition to the general issues of identifying weapons from a distance, members of the Taliban had also learned to hide them from sensors:

00:59 (Pilot): What about that guy under the north arrow, does it look like he's hold'n something across his chest?
00:59 (Sensor): Yeah it's kind of weird how they all have a cold spot on their chest.

00:59 (Pilot): Yeah it's what they've been doing lately, they wrap their *expletive* up in their man dresses so you can't PID it.

00:59 (Sensor): Yeah, just like that one, there was a shot a couple of weeks ago they were on these guys for hours and never saw them like sling a rifle but pictures we got of them blown up on the ground had all sorts of *expletive.*

Uncertainty surrounding the presence (or absence) of weapons continued, but eventually the pilots became increasingly confident that they could see rifles:

1:01 (Pilot): Yeah, I think he does have something.

1:01 (Sensor): I think so.

1:02 (Sensor): He slung it on his shoulder whatever it was, just switched arms with it or something, and is getting in the truck.

1:02 (Pilot): Alright. . . . So if the C-130 bugs out and we get a chance, dude you're pretty experienced, do what you think is right, stay with whoever you think has the best opportunity to find something on them or follow the biggest group or do whatever you want.

1:03 (Sensor): The screener is reviewing, they think something is up with that dude as well. I'll take a quick look at the SUV guys, sorry.

1:03 (Sensor): What do these dudes got, yeah I think that dude has a rifle.

1:03 (Pilot): I do too.

As the search for PID continued, a new discussion emerged about the potential presence of children in, or near, the SUV:

1:06 (Pilot): JAG25/KIRK97, your comms are weak and extremely broken uh understand we are still looking for PID, we are still eyes on the east side working on PID, we have possible weapons but no PID yet, we'll keep you updated.

1:07 (MC): Screener said at least one child near SUV.

1:07 (Sensor): Bull(expletive) . . . where!?

1:07 (Sensor): Send me a (expletive deleted) still, I don't think they have kids at this hour, I know they're shady but come on.

1:07 (Pilot): At least one child. . . . Really? Listing the MAM [military-aged male], uh that means he's guilty.

1:07 (Sensor): Well maybe a teenager but I haven't seen anything that looked that short, granted they're all grouped up here, but.

1:07 (MC): They're reviewing.

1:07 (Sensor): Yeah review that (expletive deleted) . . . why didn't he say possible child, why are they so quick to call (expletive deleted) kids but not to call a (expletive deleted) rifle.

. . .

1:09 (Sensor): Little bit of movement by the SUV. I really doubt that children call, man I really (expletive deleted) hate that.

For the next few hours, the pilots and coordinated teams observed the scene, still trying (and failing) to get a concrete PID and still questioning if children were present. They perhaps spotted an adolescent, but "teenagers can fight." Their latest estimated count was up to 20 people surrounding the three vehicles. During this stage, the weather worsened, decreasing visibility and also minimizing the amount of time the pilots would have eyes on the target (without which an airstrike would be much more difficult). However, the pilots did obtain PID when they saw a MAM pass a rifle to another MAM in the front seat of the car. After 3 hours of observing the convoy, the crew appeared to have a relatively clear view of the situation:

03:08 (Pilot): Our screeners are currently calling 21 MAMs, no female, and 2 possible children. How copy?

03:08 (JAG25): Roger. And when we say children, are we talking teenagers or toddlers?

03:08 (Sensor): I would say about twelve, not toddlers. Something more towards adolescents of teens.

03:08 (Pilot): Yeah adolescents.

03:08 (Pilot): And JAG25, Kirk97. Looks to be potential adolescents. We're thinking early teens. How copy?

03:09 (Sensor): Screener agrees. Adolescents. There's still a couple of stragglers at the other vehicles. Still upwards of 24–25 people.

03:09 (JAG25): Roger. That's our main interest now, are these vehicles and where they're heading to. We already know we have PID.

03:10 (Sensor): Pretty satisfied on just the weapons calls we made them.

03:10 (JAG25): Kirk 97, that's affirmative, from the weapons we've identified and the demographics of the individuals plus the ICOM.

03:10 (Pilot): Good, copy on that. We are with you. Our screener updated only one adolescent so that's one double digit age range.

Four hours and 13 minutes after the mission started, the Kiowa helicopters were cleared to engage the target. The air assault started at 4:16 and involved several passes over the convoy, first taking out the lead and rear SUVs and then the middle. Contrary to their expectations (that the insurgents would disperse as soon as the assault started—what they called "squirters"), after the assault began, no one ran. Instead, Sensor noted that "it looks like they're surrendering." This was not expected, and the Safety Observer on the team noted, "Dude, this is weird":

04:39 (UNKNOWN): Uh, exiting from that vehicle was probably about 4 personnel. Believe possibly two of those were female. They wore brightly colored clothing. Uh, those remaining personnel are gathered just west of the middle vehicle. They're standing about 20 meters to the west.
04:40 (MC): Screener said there weren't any women earlier.
04:40 (Sensor): Those are all people.
04:40 (MC): Yeah.
04:40 (Sensor): That's what I was worried about.
04:40 (Safety Observer): What?
04:40 (Sensor): What are those? They were in the middle vehicle?
04:40 (MC): Women and children.
04:40 (Sensor): Looks like a kid.
04:40 (Safety Observer): Yeah. The one waving the flag.

When examining how the SA developed throughout this mission, there are several errors (however, it is important to stress that we are not outlining these as errors of the pilots but, rather, a series of errors that emerged as a result of the issues of being at war and the environmental and cognitive demands placed upon such individuals). First and foremost, there were many issues with their Level 1 SA, stemming largely from technology. The pilots had to deal with changing weather conditions and often lost transmission with the ground troops and the other involved parties. The feed itself offered insufficient clarity to allow them to identify vital pieces of information (e.g., the presence of weapons), and eventually PID was not obtained until hour 4 of the observation. As such, there were several barriers in the operating environment that hindered their ability to correctly perceive important elements within the scene. For example, consider the sudden realization that there were women wearing brightly colored gowns:

04:52 (Pilot): Dude, I don't know. 'Cause we were watching these guys stop multiple times and every time they were wearing all black and only afterwards did we ever see any color.
04:52 (Safety Observer): It's possible the . . . the women and children never got out of the car, at the stops.

Hence, they were not able to see a very important piece of information while developing their SA, which affected their perception of the numbers and genders of the members of the convoy. This then, viably, also affected their interpretation of the scene (Level 2 SA). Throughout the transcript (and after receiving further intelligence from the headquarters), the pilots and observers believed they were looking at a high-level Taliban commander. This affected how they perceived the relationships between members of the convoy (attempting to identify which one was the high-level commander and who the "lookouts" were). In doing so, the "adolescents" who were identified were often viewed as military-aged, confirming their view that these were insurgents. Thus, their prior expectations (and originally established SA) altered the way they perceived the factors within the environment. In addition, this latent SA drove their search for confirmatory information (i.e., securing PID) to allow them to launch an offensive strike. As such, the lack of PID was attributed to problems with the technology, as well as hypothesizing that the insurgents were taking counter-methods to prevent PID, rather than considering the alternate explanation that there was no PID because there were no weapons in the convoy. Finally, looking at their projections into the future, given that their impoverished Level 1 SA led to a faulty comprehension of the scene, their projection as to what would happen was wrong, meaning that they were surprised by the reaction to the strikes (i.e., the members of the convoy did not run away but instead stayed where they were). It is also worth noting that this failure of SA is a feature of modern war; during trench warfare in World War I or invasions in World War II, it was exceedingly rare to misidentify civilians as combatants. Because the modern battlefield is in many ways the modern living environment, errors in identification have regrettably become more commonplace.

CONCLUSION

In combat, effective action depends on high-quality SA. However, there are many factors that can derail this process. There is uncertainty within the operating environment, but more important, even when soldiers have a good grasp of "what is going on" (i.e., strong Level 1 SA), it does not mean that they can diagnose *why* it is occurring (Level 2) or, more crucially, what is going to happen next (Level 3). This wider uncertainty often places soldiers in the position of making least-worst decisions because the uncertainty in their SA means that they are not sure if hostile actions are imminent (and hence life-saving action is required) or if what they are looking at has a benign explanation that they cannot fathom. This is often amplified by cultural unfamiliarity. SA is often developed under conditions of conflict, meaning that the decision-maker is tired, emotional, and physiologically aroused (discussed in detail in Chapter 8), further confounding these decisions. Developing SA requires extensive working memory resources, but these conditions are almost ideally situated to decrease working memory capacity through worries about safety, physiological needs, family at home, the morale of the men, and so on. When working memory capacity is decreased, our propensity to rely on heuristics increases, resulting in the application of cognitive biases, as does our proclivity to become cognitively closed (searching for information that confirms our expected SA rather than searching for information that will allow the decision-maker to establish accurate SA). As British politician

Lord Molson said, we will "look at any additional evidence to confirm the opinion to which [we] have already come." Thus, the plans we formulate are directly linked to the situation *we think* we are facing, meaning that poor SA implies the development of incorrect actions. Being "confident" in one's SA in conflict situations is incredibly difficult, and although this lack of confidence can sometimes prevent hasty actions (which is beneficial), it also has a negative consequence in that it can prevent *any* action being taken because of a constant search for information that is "good enough."

5

FORMULATING PLANS

Between two evils, choose neither; between two goods, choose both.
—Tryon Edwards

According to the SAFE-T model, once someone has correctly (or incorrectly) perceived the scene, he or she must identify the possible courses of action that are available to him or her. Plan formulation (PF) concerns how decision-makers arrive at the means to achieve their objectives, and it involves setting out various plans or courses of action that are constrained by an individual's perception of the perceived risk (van den Heuvel, Alison, & Crego, 2012). In Chapter 4, we explored situation awareness (SA), a central facet of all models of decision-making and a significant epicenter of study for those interested in decision-making under uncertainty, the role of expertise, and the etiology of error (Endsley & Garland, 2000; Klein, Calderwood, & Clinton-Cirocco, 1986; Orasanu, 1995a). However, good SA alone is not sufficient for effective decision-making. For example, Endsley (1995b) found that 26.6% of pilots' poor bad decisions occurred even with good SA. So, if a quarter of people still fail to act effectively with sufficient information, there must be other processes that can result in errors of judgment. As stated in Chapter 1, "good" decision-making requires individuals to *select* the *best* option available. In this chapter, we focus on how options are evaluated and what determines "desirability." We pay close attention to how values and goals influence decisions for both the individual and organizations and how the goals that we set can guide judgments under uncertainty. We also highlight how situations color the process through which options are generated and compared. We also discuss the inertia traps into which individuals can fall when deliberating between equally adverse outcomes.

Furthermore, we attempt to offer some insight into a philosophical conundrum at the heart of least-worst decision-making (one we have not touched upon until now). Specifically, if two options are equally unappealing, then there cannot be a "best" or "least-worst" option. If there is such an option (i.e., one option is better than the other), then the decision was not least-worst because, in theory, there was an "ideal" (or at least better) choice. In this chapter, we seek to offer some solutions to this problem by examining the process of option comparison within the individual—that is, the highly personalized and subjective experience of weighing two equally unappealing options and deciding which option, to the individual, is the best. What is important about this process, running counter to previous rational perspectives, is that option evaluation is highly individualized, and two individuals may arrive at the same choice for different reasons. Similarly, one individual may find a least-worst decision easy (a "no brainer"), whereas another individual may agonize over the choices. We expand on this in the latter part of this chapter, but first we investigate the process of comparing and selecting options.

DO WE CHOOSE?

Plan formulation, to us, is about choice. It is about having multiple different courses of action available to you and deciding which of these options you will take. Hence, PF is really in line with the prototypical definition of a decision, in that it is about selecting the best of multiple options by deliberating about consequences (Jeffrey, 1983). PF is closely tied with the concept of *option awareness* (Pfaff et al., 2013), the "perception and comprehension of the relative desirability of the available options, as well as the underlying factors, trade-offs, and tipping points that explain that desirability" (p. 155). However, not all psychologists agree that decision-making involves choice. Cohen and Lipshitz (2011), for example, "question almost every piece" of the previous definition. Recognition-primed decision-making (RPD), for example, although it involves making decisions, does not involve a "choice" between multiple alternatives (not simultaneously at least). And many psychologists have noticed that instead of "choice" being a discrete action, these distinct phases (e.g., situation assessment, option generation, option evaluation and selection, and action) occur as complex, embedded, messy, parallel cycles rather than in a clean, linear delineated sequence (Braybrooke & Lindblom, 1963; Mintzberg, Raisinghani, & Théorêt, 1976; Schön, 1983; Witte, 1972). On the other hand, our model (the SAFE-T model) and the military's doctrine (as discussed in Chapter 2) center on the process of choosing—of evaluating advantages and disadvantages of alternatives and selecting the best one. So how can we rectify this issue of choice? And, more important (acknowledging that humans make decisions with both choice and single-strategy processes), under what conditions do people *make a choice*?

Cohen and Lipshitz (2011) proposed the trimodal theory of decision-making in an effort to describe how decisions (which they viewed as "changes in commitment") occur. "Trimodal" refers to the three methods through which we can *choose* a course of action based on the degree to which we have a pre-existing commitment to a course of action.[1] The first kind of choice is to commit when there is no pre-existing commitment. In such situations, decision-makers must ask themselves, "What shall I do?" In response to this question, they will need a degree of commitment that is more specific and more comprehensive than what they already have (this is referred to as "matching). In basic terms, they must develop a course of action where none yet exists. The second form occurs when decision-makers already have a strong commitment for an action (referred to as *full* commitment). For example, as in Maj. Webb's case, a soldier may have a strong commitment to not respond to walk-ins because they pose too great a risk. In such cases, decision-makers must ask, "Will [my usual course of action] work?" If not, they must decrease their commitment to that course of action. The third form of choice (most closely related to the interests of this book) is when a decision-maker is equally committed to mutually exclusive options (*disjuncted commitment*).

The first mode of decision-making ("matching") represents following procedures, or role-based expectations. It represents what we do when there is a policy we can follow or a series of rules that dictate how someone in our role should act. Matching generates actions without choice because the choice is dictated by policy (i.e., there is no alternative). These rules frequently form an "If A, then B" model: If the situation is perceived as "A," then "B" follows without choice or competition. Although this sounds relatively mundane (merely following doctrine or policy), this kind of pre-digested decision should not be underestimated, and it can be relied on to solve incredibly high-risk complex dilemmas. In many ways, doctrine is the anticipation and resolution of decision points without the fog of war impacting upon that decision. To the degree that reliable doctrine exists, decision-makers in the moment can off-load cognition to those who came before them. If all decisions could be anticipated, and hence doctrine developed, we can envisage a world in which soldiers have no need to choose because the "correct" course of action has already been identified (via

---

[1] Cohen and Lipshitz (2011) refuted the idea of a decision as a choice between two options and instead argued that decisions are *"graded commitments of mental, affective or material resources to courses of action"* (p. 4, italics in original). Decision-making is therefore any *cognitive process* that can create, reject, or modify such commitments—regardless of the cognitive resources it may or may not consume in doing so. This, perhaps subtle, difference has many important ramifications. For example, when choice is viewed as changes of commitment, each decision is intrinsically affected by the agent's pre-existing commitments (Bratman, 1987), which in turn shape new intentions or generate immediate action (Cohen & Lipshitz, 2011). This is not a surprising revelation given that experimental research has long shown that committed decision-makers are reluctant to change their minds (the sunk cost fallacy; Arkes & Ayton, 1999), they ignore and distort information to match their currently held views (confirmation bias; Poletiek, 2001; Wason, 1968), and even deepen commitment in the face of negative feedback (escalation of commitment; Staw, 1976; Staw & Ross, 1989).

doctrine) and taught to them. Alas, this is not the case (as so many examples in this book show), yet the importance of doctrine in military decision-making should not be overlooked. To put this point in perspective, we present a decision made by one of our Special Operations interviewees (i.e., a soldier who conducts unconventional missions). As you progress through this case, consider what you would do: What are your emotions calling for you to do, and how do you want to act? Imagine the conflict you would be experiencing in making a decision and the risks you would be weighing between the different options. In doing so, you might feel a great degree of cognitive conflict (as well as accountability and anticipated regret). Add in the physiological and psychological strain of making such decisions with little or no sleep, and this situation becomes unbearable. Once you have experienced this conflict for yourself, consider the same situation from the point of view that doctrine demands only one course of action, no matter the costs, and we hope you will feel this burden of cognitive conflict lifted from you (although the emotional conflict is likely to remain).

AFGHANISTAN, LATE SEPTEMBER 2001 (PRE-INVASION)

By the time we interviewed Lt. Col. S. Green in 2015, he had completed more than 15 rotations in Afghanistan alone. He is a lieutenant colonel in the Navy but describes himself as a "Special Ops Guy." He has undergone SEAL training (the US Navy's Sea, Air, Land Team training—described by Luttrell (2007) as one of the most difficult Special Forces training programs in the world). During the past few years, he has commanded a squadron of Special Operations Forces in Afghanistan, Iraq, and the Philippines. Lt. Col. Green was in Afghanistan "a little over a month after September 11th mostly doing 'clandestine operations, small units moving alone,' or community engagement." In early November 2001, Lt. Col. Green was in one of the first convoys moving toward Kandahar, Afghanistan. They had linked up with Hamid Karzai and were spearheading the early operation into Afghanistan: "Day to day, a lot of gunfire, living very minimal, sleeping on the hoods of a vehicle on the corner of a local field. . . . So living in some pretty rough conditions." This particular convoy involved approximately 12 elite Special Operations Forces (SOF) teams, elite meaning "the premiere forces from the countries involved."

At this time, the rules of engagement meant that no SOF team could go outside the wire without a certified Joint Terminal Attack Controller (JTAC). JTACs are on the ground to help forces in the air locate targets and guide bombers toward them (usually using lasers to tag the intended target). Lt. Col. Green was the Senior Fire Support, meaning that he managed all JTAC requests in the field. So not only did he prioritize and submit requests but also he served as the advisor to ensure that munitions were directed in the right area and addressing the right targets. This role encompassed legal decision-making in war, and decisions were often made in conjunction with legal experts. Prior to this deployment, Lt. Col. Green was also overseeing JTAC training, so as he admits, he was "pretty steeped in doctrine."

On the night in question, there were five SOF teams out in the field, and most nights they had "something going on," meaning that there was kinetic contact with enemy forces and the need for close air support. This night in particular, a 12-man US Army SOF team was out on a mission to a compound which at that point was known to be occupied by insurgents. At the same time, a German team was outside the wire observing a potential threat. The Germans soon indicated to Lt. Col. Green that they had been spotted by the insurgents and wanted to know if air support was available. At the same time, the US SOF team also began to report enemy fire. It is at this point that Lt. Col. Green needed to assess "who has the more legitimate claim." We should also mention that Lt. Col. Green is conducting this support on a field radio, in the snow, at night, trying to get a good understanding of what each team is looking at.

Lt. Col. Green has to decide who his priority is because only one target area can be covered by air support. This is a classic least-worst first decision. His initial assessment was the following: What is the immediate threat and what are your options? The Army SOF team is very alarmed ("Muzzle flashes, shots being fired, we need air support now!"), and the conversation roughly goes as follows:

Lt. Col. Green: What is the target area?
US SOF: We are on a ridge. They were on a ridge overlooking the target compound that intelligence said housed insurgents, and from which shots were now being fired.
Lt. Col. Green: Ok, do you know who is in there?
US SOF: It's Taliban.

In the area were four other compounds, all of which would have been destroyed in a missile strike, which meant that it was important to also understand who was in all of the compounds rather just the target compound "because if you launch a strike against 3 insurgents, who are next to 10 civilians, you still get prosecuted." So, what Lt. Col. Green got from the US SOF team was that it had not been on station long, it believed it had a good idea of who was in the compound, but none of the indicators it was giving told Lt. Col. Green that the team had been able to fully assess the situation. Furthermore, Lt. Col. Green's primary concern was about the possibility of retreat:

We don't like to think of our militaries that way, but we move in very small forces, we are very vulnerable and we're only good as long as we're not detected and we have the element of surprise. . . . So once you are compromised, without fail, your reaction should be some kind of movement away from the threat.

So Lt. Col. Green asked the US SOF team, "Do you have clear avenues of escape?" It did. "Is there a reason you need to stay as close as you are?" There was no compelling reason beyond that this was an objective. "How did you confirm Positive Identification?" (PID). Lt. Col. Green stated that PID

> can be ascertained through several methods, including seeing a uniform, or you can have open acts of hostility [acts of aggression to you], or you can have an in extremis situation; in which there is an immediate threat to you and risk of loss of life or limbs. So when you meet one of those conditions, legally you are able to act. And not before that.

In Lt. Col. Green's assessment, these conditions had not been met: No one had been hit, the equipment had not been hit, and he knew the team had an escape avenue. The team was not able to identify insurgents (other than muzzle flashes). So, to him, PID was not being met.

At the same time, the German SOF team reports that 14 vehicles are moving toward it in a search pattern (i.e., they are trying to spot the SOF team): "Hey, they know we are here and they are looking for us. We can tell what is going on here. These guys are getting closer and we are hunkered down." So, again, Lt. Col. Green asks the German team the same series of questions: Do you have avenues of escape? Is there a reason you have continued moving forward? Have you got PID? The German SOF team reported it did not have PID, but it "knew these guys were bad just by their locale and we are really concerned."

Both teams are now requesting air support, and Lt. Col. Green has the German representative to his left and a US Army Sergeant Major to his right both demanding that support be provided to their troops. Increasing the stress, the Sergeant Major and German representative are engaged in a heated exchange regarding the apparent bias to supply air support to American SOF forces over other nations in the coalition.

As we process this decision, let us revisit the "difficulty buttons" identified in Chapter 1. We can see that this decision hits many of them:

1. *Too little information*: Lt. Col. Green cannot accurately understand what is going on, who are the targets, who is in the other compounds, and whether they are insurgents (and crucially whether there are any

non-insurgents in the area also). In short, there is not enough Level 1 SA to enable development of Levels 2 and 3.
2. *Uncertainty*: Uncertainty is high, there is poor communication, and they are relying on old equipment.
3. *Inability to diagnose the problem*: Although both US and German SOF teams are confident they have accurate SA, Lt. Col. Green cannot diagnose the problem.
4. *Multi-agency communication*: Lt. Col Green (a Navy officer) is working alongside Army units from several different nations.
5. *Emotion*: Any situation involving potential harm to troops will obviously be emotive. However, the emotional argument occurring in the headquarters about the bias of air support for American troops further increases the impact of the situation.
6. *Confidence*: It is clear that confidence in the assessments of the two SOF units on the ground is low.

Given these pressures, this decision *should* be hard. It is clearly least-worst. Support the US SOF team and potentially leave the German SOF troops surrounded by an insurgent network that is actively searching for them; give air support to the German SOF team and risk leaving the US SOF team under fire from an insurgent compound; give no one air support and risk both; or give either one air support and risk a potential error and the legal ramifications of providing air support without PID (of which Lt. Col. Green is acutely aware). However, because neither team had reached the doctrinal threshold for receiving air support, Lt. Col. Green was able (with little cognitive conflict) to be confident that the correct decision was to not give air support to either team. In this least-worst decision, Lt. Col. Green was able to make the right decision solely by following doctrines set in place surrounding the use of air support. He

> clearly told them that I had two teams with significant threats at the time and I could not pull aircraft towards them because they had not met the criteria, and I repeatedly urged them to back off the target and hit it another day or spend a couple of hours gathering further intelligence.

Thus, although the "Army were screaming for air support and the Germans were screaming for air support, neither have satisfied my criteria because they had evacuation points and no PID." Although there are aspects of his decision-making that he was not happy with in hindsight (discussed later), Lt. Col. Green maintains that he "made the right call." And, as discussed in Chapter 7, he did make the right call.

The is a clear example of *if A then B* decision-making in action. Here, Lt. Col. Green followed the rule of law—that without PID (and when the team had the option to retreat), he could not authorize an air strike. Hence, despite the presence of so many "difficulty buttons," there is little cognitive conflict. Here, we see the strengths of matching decision processes: When there is doctrine or policy, there is little room for *choice*.

Reassessment, the second form of decision-making proposed by Cohen and Lipshitz (2011), is essentially RPD. In reassessment, the decision-maker asks, "Is my course of action reliable?" (Cohen & Lipshitz, 2011, p. 10). This question can be answered by projection of the outcomes of a preference (either through storytelling or through mental future simulation; Cohen, Freeman, & Wolf, 1996; Endsley & Garland, 2000; Klein, 1998; Pennington & Hastie, 1993). Reassessment therefore closely represents the common RPD model of decision-making under risk and uncertainty in that it centers on the "workability" of a situation and accepting (or rejecting) the feasibility of the option that usually appears to the decision-maker quickly (based on prior experience). Given the overlap between reassessment and RPD, and the discussion of RPD and military decision-making in Chapter 2, we are confident that the reader is well aware of the mental processes involved in this type of choice and, importantly, the benefits and constraints of this type of decision-making.

The final form of decision-making proposed by Cohen and Lipshitz (2011) is *choice*. It begins with the commitment to choose one course of action (rather than a predisposed commitment to a course of action) and centers on the question, "Which of these options is the best means to my ends?" (p. 11). This form of choice is centered on maximizing the expected desirability of future states, and it requires shifting our commitment from multiple equally desirable (disjunctive) options to just one. As Cohen and Lipshitz argue, decisions require actual choice (rather than matching or reassessment) when

- the environment presents multiple options;
- an external pressure (organizational or cultural) requires the decision-maker to justify his or her action by comparing it to another option; and
- the goals of the decision-maker compete with one another, and advantages and disadvantages must be traded off in order to approach optimality.

DO SOLDIERS CHOOSE?

In conducting this research, one of the authors (NDS) came across a perplexing phenomenon: Some members of the Armed Forces often said that they "did not make any decisions" on their deployment. Although originally this was difficult to reconcile (they had often been deployed in combat roles and engaged in kinetic activities, yet felt like they made no decisions?), perhaps it is explainable in Cohen and Lipshitz's (2011) framework. Specifically, what they are saying is not that they did not make any decisions; they made many decisions daily. What they are saying is that they never made any *choices*.

From this, we can infer that in war, many decisions do not specifically require a choice. Actions are either mandated (through an SOP) or there is an analogy or preexisting commitment that (at least according to forecasts and current SA) will work. This assertion is fine, and indeed it is the basis of military training and really does underscore the importance of RPD in both training and theory. As Mike Matthews (2013), a professor at the US Military Academy (West Point), highlights, continual practice leads to a "library of schema–script connections. As proficiency builds, the soldier can quickly pattern-match the observed situation with an appropriate script. Decision making in these cases may become virtually automatic" (p. 62). The learner therefore builds his or her own library of experiences that can be tapped into when an intuitive decision is called for (Matthews, 2013). Military decision-making research also follows this trend, focusing on RPD perspectives of how experience and expertise are applied in the field (Bryant, 2002; Elliott, 2005; Morrison, Kelly, & Hutchins, 1996; Pascual & Henderson, 1997; Ross, Klein, Thunholm, Schmitt, & Baxter, 2004). Previous experiences, a wealth of training, and SOPs therefore combine to allow the soldier, in many conditions, to act without deliberating between options but, rather, by choosing one solution (experience- or SOP-based) and acting. This is the RPD model. However, there are cases in which this will not be enough. In situations with competing goals and multiple options, soldiers will have to choose. And this choice, deliberating among the consequences of multiple options, can often cause conflict.

Perhaps one of our interviewees phrased it best:

> I always fear making a wrong decision, especially in a deliberate process like this. In terms of a right or wrong decision, if it is time critical and I have got to make a decision or else I'm going to die or somebody else is going to die, those decisions get made very easily and there's no real thinking about it in the time that you have. Here, in the 5 hours that I had to dither and ruminate and the number of "what ifs" and "if I make this decision this is what's gonna happen" the number of scenarios that I was able to go through in my

own mind were significant enough to where I did feel that making the wrong decision was going to end up in mission failure and in my own loss of credibility to in the eyes of my commander, in the eyes of my soul, and my air men and my folks that rely upon me to make those right decisions, and so I would be lying to say that . . . my thinking about making the right or wrong decision didn't affect me making the ultimate decision. But with the amount of time that I had and the amount of time that I had to ruminate on the decision I did feel like I made the right decision and I feel like when everybody looks from the outside that the end is achieved, that that was the right decision to do it. So I wouldn't necessarily say that making the wrong decision was necessarily something that I really thought about umm. . . . But what I did do was I stressed heavily about the idea that I needed to make the decision that was going to result in mission success versus a decision that was going to result in mission failure umm. . . . Again accepting the risks that were going to be accepted after that decision which is again why you have the two critical options between bad and not so bad type decision making.

Thus, despite the prevalence of RPD (both in the real world and in training), we hold that soldiers are required to make choices. They will have to evaluate options and choose an alternative based on consequence. We know that making such choices is demanding, given the nature of modern warfare and the operating environment. Furthermore, if research and practice remain focused on RPD and decision-making without choice, when soldiers face a situation that does require choice (because of the parameters set out by Cohen and Lipshitz [2011]), they may be ill-equipped to do so. This point cannot be emphasized enough.

CHOICES AND CONFLICT

Choice breeds trade-off, and trade-offs result in conflict because "decision makers must accept less of one choice attribute to get more of another" (Luce, Payne, & Bettman, 2001, p. 86). Decision-makers experience intense conflict when faced with "simultaneous opposing tendencies within the individual to accept or reject a given course of action," the consequences of which involve "hesitation, vacillation, feelings of uncertainty, and signs of acute emotional stress" (Janis & Mann, 1977, p. 46; see also Tversky & Shafir, 1992). Furthermore, more often this conflict is experienced when making a choice and not when executing it. For example, the greatest level of stress for a parachute jumper is not during the ascent nor prior to the jump but, rather, at the point at which he or she made the original decision to participate (Epstein & Fenz, 1965). As Tversky and Shafir (1992) wryly note, the "experience of conflict is the price one pays for the freedom to choose" (p. 358). In decision-making, "conflict" has no formal definition, but the most commonly held view is that it involves "preference uncertainty" (Dhar, 1997; Fischer, Jia, & Luce, 2000; Shafir, Simonson, & Tversky, 1993). Hence, when a decision involves multiple options, conflict refers to the uncertainty about which option is more valuable (Fischer, Jia, et al., 2000; Fischer, Luce, & Jia, 2000).

It is worth being explicit that conflict is not necessarily associated with the number of choices (although much research shows that more options can hinder decision-making). Instead, conflict often stems from the *nature* of the decision. Simple binary decisions can be incredibly taxing given equally negative outcomes. Capt. West (who is discussed in detail in Chapter 8), here discusses how binary decisions in the field created significant conflict:

> [During ranger training] I remember coming off the back of the helicopter in the middle of the night and I had no idea what the hell I was doing. Buildings were getting cleared, people were sending me all these radio calls through both ears, I was literally just standing there and I did not know what to say because things were going so fast. So you go 0 to 100 and you have to know how to control all these things and still understand what the tactical situation is. You're sending reports back to the battalion, you are controlling an aircraft, you are getting the clearing of a building, how many detainees do we have, I mean there is so much going on, by the time we deal with something like this, what is miserable about this [operation in Afghanistan] is that with the rangers [in training] you are in the middle of the night, you have everything, the deck is stacked. This is daylight, single file, big marshmallow men walking around the middle of a village and you are just getting pummeled. So, it is fewer things to think about, but those few things are just horrible.

Conflict is an aversive experience, and people are motivated to avoid it (Lewin, 1951). Experiencing conflict can have both behavioral (e.g., longer choice times) and mental outcomes (e.g., lower confidence). Participants in experimental studies preferentially select a "no-choice" option when the choices generate conflict (Dhar, 1997; Dhar & Nowlis, 1999). Under conditions of conflict, decision-makers take more time (Fischer, Luce, & Jia, 2000), report increased decision difficulty (Scholten, 2002; Scholten & Sherman, 2006), have lower confidence in their choice (Zakay, 1985; Zakay & Tsal, 1993), are more likely to defer (Dhar & Nowlis, 1999; Dhar & Simonson, 2003), and have an increased number of thoughts or justifications used for each option (Dhar, 1997). In our research, conflict was often accompanied by increased experience of negative affect while making a decision.

We experience a sense of conflict when we have to make both large and small trade-offs in decisions (Scholten & Sherman, 2006), but the degree of conflict one experiences is related to the *size* of the trade-offs between the attributes that one must make (Scholten, 2002; Shafir et al., 1993; Simonson & Tversky, 1992; Tversky & Simonson, 1993). To give you a sense of high and low trade-off conflict, let us consider a hostage dilemma that was posed to one of the authors (Dr. Shortland) by another co-author (Prof. Alison) during one of our (many) book-writing workshops.

THE HOSTAGE SCENARIO

Imagine you are a commanding officer and have arrived at the scene of a hostage siege at a local junior school (the students are aged between 5 and 12 years). You have received intelligence that although the majority of teachers and pupils have been able to escape the school and are safe in the car park, there are two rooms in which hostages are being held. Intelligence suggests that each room contains one hostage-taker. In the first room, from what you can gather, the hostages include two children (aged 8 and 11 years) and an elderly teacher. In the second room, the hostages are three children (aged 6, 8, and 12 years). From your assessments, you can launch a raid on one of the two rooms or split your men to launch simultaneous raids on both rooms. Launching a raid on a single room will likely result in the successful rescue of the children inside but will also very likely cause the death of the hostages in the second room. Splitting your forces would create a much lower chance of successfully rescuing the hostages in each room.

In this scenario, there are both small and large trade-offs to be made, all of which induce conflict. For example, if you decide to split your forces and launch a simultaneous raid, you would, in essence, be trading away the likelihood of success but avoiding the conflict associated with choosing between rooms. This is clearly a large trade-off that requires a high degree of sacrifice. On the other hand, if you decide that you do want to launch a raid on only one of the rooms, then you are faced with deciding which room to target. Both rooms have three people inside, hence the trade-off is relatively low, yet conflict still emerges.

Thus, conflict occurs in both high and low trade-off decisions because in high trade-off situations, conflict arises from the *sacrifices* that have to be made, whereas in the low trade-off situations, conflict emerges from having to find a strong *argument* between two similar options (Scholten & Sherman, 2006). With reference to the hostage scenario, in the decision to target a single room, sacrifice is high (i.e., you are potentially condemning the second room to death) but argumentation is low (you have a strong argument in that you can guarantee with greater confidence that you will rescue at least three hostages). Having made that choice, then deciding which room to target leads to low sacrifice and high argumentation (i.e., it is very difficult to argue between the value of lives: two children and one adult vs. three children).

Scholten and Sherman (2006) formalized this relationship between argumentation and sacrifice in their double-mediation model of conflict. In the double-mediation model, they propose two important hypotheses. The first, which we discussed previously, is that both high and low trade-offs cause conflict. The second, and perhaps more important, hypothesis is that "when one attribute is much more important that the other (very unequal attribute importance), the relation between trade-off size and conflict will change in an upward direction, more specifically, the inverse U-shaped relation will change into a (more) positive relation" (p. 243). Thus, the degree of conflict experienced emerges from three pressures:

1. The degree of *sacrifice* incurred by selecting one option
2. The ability to *argument* and justify a choice
3. An independent metric of attribute *importance*

As Scholten and Sherman (2006) outlined in their model (and empirically tested in the lab), when one attribute is valued more highly than others, the amount of conflict is "differentially weighted," meaning that the degree of conflict experienced from sacrifices and argumentation is not equal but, rather, mediated by the degree of preference we have for a given choice. Consider your reaction to the hostage dilemma presented previously: If you view attempting to save the lives of all children as far more important than successfully saving the lives of three (and in doing so not trying to save the lives of the remaining hostages), then the decision may be relatively easy because to sacrifice trying to save all lives is simply not acceptable. On the other hand, if you view successfully saving any hostages as the most important variable, then sacrificing attempts to save the remaining hostages is tolerable because you are increasing the likelihood that you will save some hostages.

What is interesting is how the double-mediation model seems to ideally integrate with our own naturalistic research on decision-makers who become inert and those who do not. In her PhD thesis on decision inertia in critical incident decision-making, Nicola Power (2016) conducted a series of critical decision method interviews (similar to those we used here) with members of the "blue light" services (i.e., Police Service, Fire and Rescue Service, and Ambulance Service). Her work with Prof. Alison focused on exploring how different goals facilitate action (and under what conditions these goals inhibit action). As Power outlined, whereas approach goals ("I want to make this situation better") influence tendencies to take positive action toward a positive stimulus, avoidance goals ("I don't want to make things worse") encourage individuals to avoid negative effects by moving away from a negative stimulus (Elliot, 2006; Elliot, Eder, & Harmon-Jones, 2013). Power interviewed 31 command-level decision-makers from the blue light services. Similar to our own experimental method, participants in her research were asked to recall a "difficult decision" that they had responded to in the past. Power found that emergency commanders held two overarching goals:

1. Save life: Goals and motivations associated with *approaching* positive outcomes from a situation
2. Prevent further harm: Goals and motivations associated with *avoiding* anticipated negative consequences

However, Power found that these (competing) goals often resulted in uncertainty, goal conflict, passive and active avoidance, and inaction. In her words, "The 'save life' goal appeared to derail action if the decision maker experienced goal conflict by trading it off against the competing avoidant goal to 'prevent further harm'" (p. 96). In terms of the double-mediation model of conflict, emergency service responders experienced significant conflict because they were unable to trade off between the goal to save life and the goal to avoid further harm. Referencing the important third variable of attribute importance: They had no preference. As one of her interviewees recalled, "What you have to avoid is delaying making your decision about anything which then leads to somebody getting hurt but by the same token you don't want to knee-jerk and rush into a decision that is not properly considered" (p. 96). Thus, because the service personnel viewed these goals as equal (and hence competing), they experienced significant conflict and, accordingly, experienced decisional conflict resulting in decision avoidance.

In military situations, although competing goals are common, we found that goal conflict was far less prevalent. In line with the double-mediation model, one viable explanation (and one that we hold to be true) is that members of the Armed Forces have a strong sense of attribute differentiation when choosing between least-worst decisions. In Power's (2016) work, members of the blue light services encountered decision inertia when they were required to make a trade-off between approach and avoidance goals, both of which were equally important to them. Soldiers, on the other hand, in our research, consistently demonstrated a strong degree of attribute differentiation between different goals (meaning that they often had a strong goal hierarchy), and when making least-worst decisions, they often would not make a sacrifice on a single prominent attribute. Namely, they often refused to make any sacrifices regarding the safety of the men and women under their command. For example, when deciding if and how to recover a large military asset currently burning in a local village, one interviewee highlights the importance of force protection:

> I think um If I was to prioritize [my goals], it would be to protect the lives of US Armed Forces, I don't want to be skittish and afraid to do our mission, but I certainly didn't want to waste lives, or put them unduly at risk. We had spent a lot of money to mitigate risk and injury and death, and this was about recovering a piece of equipment, and I put that below soldier safety. If I would have known the outcome of that would have been the death of 2 soldiers, I would have said it was not worth it.

On the other hand, his decision-making would have been completely different had there been military personnel at risk:

> If I had personnel out there with it. I mean say that the truck that they were hauling on was you know, say there was some personnel trapped out there. And then those are the times when you know you have to go, and you put people at risk to go, and we've been through that a lot of times when people get hit by an IED [improvised explosive device] and there are folks trapped in the vehicle and then you just go, you don't think about it, you don't feel like you have a choice to just leave people out there.

Similarly, another participant highlights the importance of force protection when deciding how to proceed with the convoy:

> Of course, [force] protection is number one, make sure everyone gets back in one piece, hence we have a lot better chance in the firefight. I guess everyone in the country has control, and we have a squad that has more fire power than one of the most fire powered squads in the military so the special forces or what . . . our job is personal security detail, we are there to make sure an officer made it back and that was our primary objective, everything else was secondary.

Another interviewee also emphasizes the central importance of protecting soldiers in all the decisions he made:

> You never know if it is the right decision or not. Someone could have parked a car bomb right there on the side of the street next to it, you could have had

multiple casualties, stuff like that was running through my mind. Stuff like that was running through my mind you know, what is the worst thing that can happen. You see it might be, plugging the cobalt beneath the street full of explosives and just waiting for us, and just baiting us with that. And all that stuff was certainly in the back of my mind. It was those kinds of risks. And then someone would say, man that was totally the wrong call, putting that many people in that kind of an area, you just created this huge target. And you know, was this scraper worth it? So that was always there. But I always felt like, and I probably felt on most of my missions that I always did everything I could do to mitigate risk.

A captain we interviewed from the US Army shared a similar sentiment:

I mean the idea is the way I saw it, and the, it is, you know, I, uh, before every deployment I talk with parents and, you know, my mission is to do my best to bring every one of their sons home.

As one of our interviewees highlights, the need to protect all of one's soldiers stays even though it often clashes with more mission-focused objectives:

You don't want your squadron commander to yell at you. So you don't want that. So I mean that was of course in the back of my mind, you don't want to fail your mission and then again you don't want to lose all of [vehicles] so that was a stressor. So you didn't want to fail your mission ... certainly didn't want to lose a soldier.

The previous collection of quotes give a sense of the overarching importance of force protection that flowed through all of our interviewees and their decision-making calculus. Thus, despite the many goals in military decision-making (protecting the population, protecting forces, and achieving the mission), because these attributes are weighed so differently, decision conflict is rare.

Perhaps one of our Marine interviewees sums it up best when, while pursuing a high-ranking member of Al-Qa'ida in Iraq, he maintains that force protection was his number one priority:

Targeting this insurgent was an incredible, great, opportunity, we have never had an opportunity like this before to get some bad guys that have put up a fight. But my goals were; (1) Protection of own force; (2) Killing bad guys; (3) Protecting population.

The importance placed on looking after one's fellow soldiers is not surprising: It is commonly known to any layperson that this strong social bond between soldiers is a vital protective factor from the stressors placed upon them during war. In Wong, Kolditz, Millen, and Potter's (2003) study on how soldiers were motivated to "continue in battle, to face extreme danger, and to risk their lives in accomplishing the mission," they found that "U.S. Soldiers, much like Soldiers of the past, fight for each other" (p. iii). Thus, today's soldiers are similar to their World War II counterparts, for whom Marshall (1947, p. 42) noted,

> I hold it to be one of the simplest truths that the thing which enables an infantry Soldier to keep going with his weapons is the near presence or presumed presence of a comrade.... He is sustained by his fellows primarily.

He continued: "Men do not fight for a cause but because they do not want to let their comrades down."

SACRED VALUES

The previous discussion implies that protecting one's troops confers psychological coping benefits; such strong group ties have psychological consequences for how soldiers are able to handle conflict and navigate least-worst decisions. To us, the central importance of this factor implies that it is not just an "important attribute" in decision-making but also something more ethereal or spiritual. For most military personnel who we interviewed, "force protection" was more than an organizational or cultural norm; it reflected a "sacred value."

Values are types of beliefs that guide us toward value-congruent behavior (Bardi & Schwartz, 2003; Schwartz, 2005, 2009). Values affect the trade-offs we are and are not willing to accept in decision-making (Kruglanski & Stroebe, 2005). It follows that incorporating values into decision-making can increase decision difficulty (especially when there is value conflict) *or* can facilitate decision-making because people often hold values that are absolute—that is, they are precluded from being traded off or traded against (Hanselmann & Tanner, 2008). Inviolable values are called "sacred values" (Tetlock, Kristel, Elson, Lerner, & Green, 2000). Sacred values are defined as "any value that a moral community implicitly or explicitly treats as possessing infinite or transcendental significance that precludes comparisons, trade-offs, or indeed any other mingling with bounded or secular values" (Tetlock et al., 2000, p. 853). Baron and Spranca (1997) referred to such values as "protected," in that each protected value is "infinitely more important than others" (p. 2), and attempting to trade off against such values can elicit strong emotional reactions such as denial, blame, procrastination, and avoidance (Anderson, 2003; Fiske & Tetlock, 1997).

Sacred and secular values in decision-making lead to three distinct types of value trade-off (Tetlock et al., 2000): First, *a routine trade-off* in which two secular values are pitted against each other. In the case of the hostage rescue, those secular values (something important but not sacred) may include the desire for success with organizational policies. These decisions are hard, but mainly because we do not care about either variable *enough*. The second type of trade-off is the *taboo trade-off*, in which a secular value is traded off against a sacred value. Let us say, for example, that you hold that you must "try to save every life" as a sacred value. In this instance, this sacred value is traded off against the more secular value (likelihood of success), resulting in a difficult but consistent decision that you will launch a simultaneous raid because that way you will not simply ignore your sacred value of trying to save lives. Taboo trade-offs are "in this sense, morally corrosive" in that "the longer one contemplates indecent proposals, the more irreparably one compromises one's moral identity. To compare is to destroy" (Tetlock et al., 2000, p. 853). The mere contemplation of a taboo trade-off is sufficient to elicit a strong negative feeling of distress (Tetlock, 2003). The final type is a *tragic trade-off*. Considering the hostage example, there are several values at play that may be sacred. For example, the innocence of children, the sanctity of any life, and the need to ensure that you tried to save all possible lives could all be sacred. In this case, the least-worst decision is a tragic trade-off because it requires individuals to trade two sacred values that ordinarily would both receive absolute priority.

Perhaps it is easiest to present this as a simple table (Table 5.1; we acknowledge the oversimplicity of this representation) in which options A and B can be either sacred or secular. To complete the table, we add our observations of the likelihood that inertia will emerge in these instances and the ability of the decision-maker to make fast and effective decisions.

So, in least-worst decisions, values (especially sacred values) can both facilitate and hinder effective decision-making. It is clear how two competing sacred values can create a toxic decision environment in which the decision-maker is forced to sacrifice something that he or she holds sacred. Such is the cost of holding something to be sacred. Previously, we made the point that sacred values can make difficult decisions easier when, all else held equal, there is one non-negotiable value. To put this point in perspective, we present the case of Capt. Scott, who, in his words, made a decision that he knew was "the end of [his] 10 year career." Now, in normal situations, the decision to throw away a 10-year career would not be made lightly, but here, guided by a sacred value, Capt. Scott made this decision with great ease and great speed.

Table 5.1 Sacred and Secular Values and Least-Worst Decision-Making

		Option B	
		Secular Value	Sacred Value
		Option A	
Option A	Secular Value	*Routine trade-off* Decision inertia due to the inability to decide between two opposing secular values (e.g., organizational policies)	*Taboo trade-off* Effective decision-making /absence of decision inertia due to a sacred value (e.g., force protection) that cannot be traded against
	Sacred Value	[Identical to top-right conditions]	*Toxic trade-off* Decision inertia/extreme difficulty in effective decisions due to trade-off between equal sacred values that decision-maker is not willing to trade off against

"I'M NOT ASKING YOU. I'M GIVING YOU AN ORDER"

Capt. Scott was a company commander in Afghanistan between 2003 and 2005. His company was running security force missions for the civil affairs team. The company would accompany them into the Tora Bora region of Afghanistan and provide security while the doctors and dentists performed basic procedures for members of the local villages. Capt. Scott sent a platoon out (along with 30–40 civilian affairs personnel) to do a routine mission up in the Tora Bora to examine construction progress for a medical clinic. Capt. Scott stayed back to provide force protection for the forward operating base. At approximately 0200, the deployed platoon was hit by Taliban fire: "Umm just a bad situation. High ground all around err and they got up there and about 2 o'clock in the morning they got hit. We found out later it was about 200 Taliban." If not for a close air support mission, they "probably would have all died out there on that rocky knoll." But air support enabled them to stave off the fight long enough to call back for ground support:

> Because by the time they got done probably about a 45-minute fire fight . . . they were critically low on ammunition, clearly outnumbered and the only thing they had going for them was close air support which probably again saved their lives.

Thus, this left Capt. Scott with a least-worst decision:

So, the first decision that I had to make, which I think was fairly easy was do I use my quick reaction force [QRF] to go up there? This would then leave the base protected by about a platoon plus the other military folks to support the people that where there. So, do I use my QRF to go up there and get them and reinforce them and bring them back, knowing that there was one way in and one way out so we were probably gonna get hit on the way up or most certainly on the way back home? Or do I leave them there and stay back to protect the base because you know those could be a ruse to draw us out and then hit the base?

Ultimately, Capt. Scott decided to activate the QRF. It took them approximately 90 minutes to get up there—it was not far, but there were no roads and they were running on goat trails. Their arrival allowed them to reinforce the troops in contact. Here is Capt. Scott's second decision point:

I'm sitting there getting everybody consolidated, reorganized, and distributing ammo because we didn't know if there was gonna be another attack or what was going on and we had the civil affairs major came up to me and he said "I want you to pursue the enemy. I want you to go up into the mountains and get a body count pursue the enemy," and I looked at him and I said "Are you kidding me?" And, of course, we were all standing around the top of this little hill mountain that we were on where this medical clinic was no bigger than this room right now and we were in the middle and everybody was kinda around getting stuff ready and you know . . . I said "We're not gonna do that. We're gonna consolidate and reorganize, we're gonna distribute ammo and we're gonna get the hell out of here. We still have to make it out of this canyon. They let us in but it doesn't mean that they're gonna let us out. And frankly we don't have the force to pursue the enemy." I mean at the time we didn't know how big they were, we didn't get the intel reports and stuff back until later, but it was clear that they were outnumbered and damn near overrun but for the close air support. And he said "Captain I'm not asking you, I'm giving you an order you will go into the mountains and get a body count and pursue the enemy."

Captain Scott's values clearly drove his decision-making in this instance:

Talk about paralysis by analysis, there was no paralysis here. For a split second I thought "This is it, this is my military career" . . . and I said "Well, Sir here's what's gonna happen; the people that are wearing this patch" [pointing at his own arm]—which was everybody there expect for the civil affairs team which were probably 6 guys—I said "Everybody that is wearing this patch is gonna get in their vehicles and we're gonna get the fuck out of here because this is not a safe place to be. And the people wearing that patch [pointing at the civil affairs Major's arm] can stay here and follow you into the mountains." And at which point he got right in my face, very irate. the guy had a temper problem. He was just extremely irate you know swearing at me telling me that I was disobeying a direct order and he was gonna bring me up on charges and I just looked at him and said "Well, you do what you have to do Sir and I'll do what I have to do." And I looked over to my Platoon Leader and I said "Mount up let's get the fuck out of here" and everybody got in their vehicles including the Major who was the civil affairs major and we rolled out of there.

Now this clearly presents a least-worst decision: Leave the base unmanned or leave troops in contact without support. However, depending on the value system of the soldier, this decision could be relatively easy to make (and in reality, it was). Leaving the base unmanned may represent a secular value (i.e., it is important, but there is no moral or spiritual obligation to it), whereas leaving troops in contact (and hence at risk of harm) would violate the sacred principle of protecting fellow soldiers. In this case, we are presented with a routine taboo trade-off. Theory tells us the sacred value will prevail, which is exactly what happened. However, this value system was further tested once they had successfully helped the platoon because the civilian affairs major approached our decision-maker with the order to "pursue the enemy. I want you to go up into the mountains, get a body count and pursue the enemy." This again presents a further least-worst decision: head into the mountains with a tired (and underpowered) platoon to pursue (what turned out to be) more than 200 Taliban forces or disobey a direct order by turning around and returning to the forward operating base. Once again, our soldier was able to make his decision quickly and with relative ease:

My decision-making process there took about 30 seconds. Long enough for me to think . . . literally I disobey this order I'm taking my 10-year career and throwing it in the garbage. But again, this was a split second decision for me because at the end of the day I remember thinking to myself I would rather lose my commission, you know, be fired and find another job and have everybody there that was with me make it back home than you know make the wrong decision and follow an order that I knew was tactically unsound and lose my soldiers and/or my life. So, I don't think that process took very long I guess long enough for me to kind of kinda have that conversation in my mind.

This soldier demonstrates the sacred value of force protection over the secular value of obeying orders to superior commanders and is able, in an extreme and high-consequence situation, to make a fast and effective decision with little to no cognitive conflict. His following explanation further emphasizes our point:

> No you know I ... it sounds corny but I think there's a warrior ethos. There's this idea that you never leave a fallen comrade and I think that influenced heavily in my decision making. My radio telephone operator [RTO] and I had a conversation when we went into country and it was right after one of our soldiers got dragged away in a mission. He got captured and you know they ended up either decapitating him online or they did something like that, and I remember looking at my RTO and saying "Hey no matter what, don't let them take me alive" and vice versa right. I mean and we were serious like if there is nothing else put a bullet in me I just don't want to end up on CNN right so ... I think that idea that you never leave a fallen comrade is really engrained at least in me and I think a lot of my fellow comrades and I think that has a lot to do with the decision-making process there was almost no doubt I was going to get those guys right? In my mind there was almost no doubt so ... for what it's worth I don't know if that's helpful to you.

Perhaps this quote, more than anything else you will read in this chapter, emphasizes the unwavering importance of force protection in our soldier sample.

Although there is extensive research on both taboo trade-offs and decision-making (e.g., Duc, Hanselmann, Boesiger, & Tanner, 2013; Hanselmann & Tanner, 2008; Tetlock, 2003; Tetlock et al., 2000), and moral decision-making as it pertains to military personnel (e.g., Eriksen, 2010; Hartle, 1989), none has gone so far as to investigate the effects of sacred values or taboo or tragic trade-offs in military decision-making (especially as it pertains to the emergence of decision inertia). However, using our own data and those from other naturalistic research in this area (e.g., Power, 2016), we are able to demonstrate the role that sacred values can play in least-worst decision-making. Furthermore, this view explains why certain individuals may struggle with some decisions, whereas others will not. Consider the risk of civilian casualties in Afghanistan. General Stanley McChrystal, then Commander of the US Armed Forces in Afghanistan, concluded in his 2009 military progress report that there was an urgent need for a significant change in the way that the International Security Assistance Force (ISAF) was operating in Afghanistan. In his words, ISAF "had shot an amazing number of people, but to my knowledge, none has ever proven to be a threat" (McChrystal, 2009; Oppel, 2010, p. 3). ISAF-caused civilian casualties were eroding ISAF's credibility among the Afghanistan population and, in so doing, significantly bolstering the Taliban's strategic goals. When Gen. McChrystal took over as Commander ISAF (COMISAF), he implemented a series of strategic and tactical innovations and adaptations focused on minimizing civilian casualties through more restrictive rules of engagement, increased alignment of civilian and military efforts, building up the Afghanistan National Security Forces (ANSF), and emphasizing nonkinetic activities. The goal was not to eliminate kinetic operations but, rather, to ensure that kinetic activities were deployed in a counter-insurgency-centric manner that emphasized protecting the civilian population as the ultimate target. As such, minimizing civilian casualties became a more pressing organizational policy, and the data show that this did lead to a direct reduction in the number of civilians killed by ISAF forces (Shortland & Bohannon, 2014). However, not all soldiers adopted this new policy. Consider, for example, when one of our interviewees had to launch a missile attack on a compound: He "knew what the collateral damage estimate was, I was just going to have to accept it." Contrast this with Maj. Webb (whose story we told in Chapter 4, "The Walk-In"), who, while holding that saving his fellow soldiers is paramount, viewed avoiding civilian harm as equally important:

> But there is the other side of me that looks at it and says, you know, that if I fired and wasn't supposed to, not only would that have probably ended my career, forget my career, I would have had to have lived the rest of my life knowing I had killed that guy. I don't know what other people's impression of the military is, if we take these decisions lightly, but I certainly didn't and I don't think other people do either. And the idea that you become jaded to the point when you stop caring about hurting innocent people. I can't imagine becoming that jaded about that decision, in fact it still bothers me to this day, the idea that I, in a fraction of a second, I could have shot that guy, and if he was you know, Taliban, great, but if he wasn't, you know, I can't imagine the thoughts that would go through my head about that guy's family, and things like that. So, it is a hard decision to live with.

Thus, individual differences in sacred values can upend the decision-making calculus and, crucially, explain when and why some individuals may suffer indecision and inertia, whereas others will not. It is not about the objective attributes of the choices but, rather, the interaction of the choices on offer and the importance of the values that each choice would sacrifice against.

To further emphasize the presence of individual differences in sacred values, we mention that the sacred value of protecting one's fellow soldiers, even though it was highly important for all, was not always viewed as the most important value. For example, Capt. West (who is discussed in detail in Chapter 8) navigated several least-worst decisions throughout a deployed mission in Afghanistan, but his sacred value was completing the mission, not force protection:

Our duty is not to bring everyone home. As a young lieutenant, I thought that my duty is to bring everyone home safe, but that is actually not what I promised to do. I promised to execute what, from way up there, you're advancing US policy and objectives, but at my level it is executing missions. Now, at that point it was "I don't think I can complete this mission with this group of folks that I have. I don't think we can get it done." So, I don't know if it goes deep enough as to why, I am objective focused, but I think largely it is because I try my best to place as much trust in my leadership to not put me into situations where we are just getting screwed.

Perhaps as readers work through Chapter 8, they might consider how this paramount focus on the mission influenced the decision-making of Capt. West.

CONCLUSION

A hungry donkey stands between two identical hay piles. The donkey always chooses whichever hay is closest to him. Both piles are exactly the same distance apart, one on his right, one on his left, and they are identical in every way. Which pile of hay will the donkey choose to eat?

This paradox, originally proposed by Aristotle ("A man, being just as hungry as thirsty, and placed in between food and drink, must necessarily remain where he is and starve to death"), was made famous by the French Philosopher Jean Buridan as a central argument against free will. According to this paradox, the donkey will be unable to choose and will, in turn, starve to death in the paralysis of choice. Although commonly used in political satire, this philosophical conundrum reflects a common observation that when people face equally attractive (or unattractive) choices, they can become paralyzed. This issue of being unable to choose between equally unappealing options has been central to this book. However, in line with the previous argument, if a choice is truly least-worst, how can any decision ever be made? In this chapter, we sought to address this paradox and explain how, although on the face of it a choice may be least-worst, people can make a choice.

We started this chapter by investigating the nature of different types of decisions. We outlined the role of doctrine, policy, and previous experience in removing the need for choice. However, although we can eliminate the need for choice to a degree (which is the central ethos of military training and decision-making theory—to better equip soldiers to make fast, intuitive decisions with little deliberation), we cannot eliminate it completely. Choice breeds conflict because choosing requires sacrifice. We outlined the emergence of cognitive conflict and its consequences. In our own work, we have found that when faced with trade-offs, decision-making often stalls and the decision-maker struggles to commit to a choice. But here, from what we have seen, members of the Armed Forces are better able to handle conflict and often less vulnerable to decision inertia. In this chapter, we sought to explain this by examining value systems. In trade-offs, the degree of conflict experienced is weighted by the pre-existing values we hold, and some of these values we will not trade off against (sacred values). In the military, the bond between soldiers is well known, and it is an important protective factor for the psychological struggle of being at war. But here, we extended the importance of this, showing that in many cases, the sacred value of force protection (or any other sacred value, such as completing the mission) often drove decision-making under conditions of conflict because soldiers refused to trade off against this, meaning that they were able to make fast and effective decisions.

Thus, with reference to Buridan's ass, which one he chooses, in fact whether he chooses *at all*, comes down to his ability to identify which one he cares more about—being fed or being hydrated.

6

EXECUTING PLANS

Five frogs are sitting on a log. Four decide to jump off. How many are left?
Answer: Five.
Why? Because there's a difference between deciding and doing.
—Feldman and Spratt (1998, p. 1)

This book has so far focused on the process through which choices are made. But, just as the frogs on the log, choosing is only half the battle; after making a choice, one must actually commit to the actions that enact that choice. In this chapter, we discuss the plan execution stage of the SAFE-T model and the role of the wider response team in turning a choice into a behavior and (it is hoped) the intended and ideal outcome. We focus on this because, by their very nature, least-worst decisions create internal conflict, but this conflict also exists between members of a responding team. This is especially true if different members of a team have different value systems. Here, we investigate what such conflicts can look like and, critically, how decision-making can be protected when conflict emerges. Specifically, we focus on the role of trust—how it is formed, both in short- and long-term interactions, and the benefit it provides to decision-makers.

The definition forwarded at the start of this book described a decision as a "commitment to a course of action." That means not only are decisions assessments of what to do and how to do it but also they encompass *doing*. Furthermore, although a decision may be an internal commitment or preference, in many cases the decisions made by an individual are made within an environment of multiple, competing individuals and agencies that may (or may not) share that individual's priorities and values. As such, in many situations, the commitment made by the individual may not reflect the ultimate decision. In this chapter, we examine the wider social pressures present within the decision-making environment (ignored in both recognition-primed decision-making and military and traditional perspectives on decision-making) and consider how these can interfere with an individual's preference, resulting in changes in commitment and decision errors. We also examine barriers to executing decisions, such as competing values, paying special attention to instances in which the decision-maker's values clash with those of members of the team(s) that they operate within and alongside.

Let us start with a callback to an account we presented in Chapter 5. As the reader will remember, Lt. Col. Green was placed in a situation in which he had to choose to which Special Operations Forces (SOF) team (US or German) he would provide air support. To review, both the US and German SOF troops had determined that they had come into contact with insurgent forces, and both believed they had sufficient identification to warrant air support. Lt. Col. Green disagreed: Given that neither group had obtained positive identification (PID), and that both had a possible avenue of escape, Lt. Col. Green thought neither side met his criteria for air support. His decision (based on current information) was to put air support in a holding pattern between the two SOF units (so that if anything changed, air support could be deployed to support either unit). In his mind, he made a decision to not provide air support. However, as the reader will see in the following account, others' competing values and priorities led to an outcome at odds with Lt. Col Green's decision.

TROOPS IN CONTACT

Lt. Col. Green highlights that as he deliberated, the Sergeant Major for the US Army was "extremely vocal and yelling." Despite this, Lt. Col. Green kept reiterating to the Sergeant Major that they had not met his criteria and he was not providing support. He told the Germans the same thing. So, his assessment was that "both of the teams had [potential] legitimate threats, but both had avenues of escape, and if I moved my resources to the wrong one they're split by large distances and then I've really compromised myself." The issue, however, was that the situation then escalated as the US forces declared that they needed aircraft *now* because gunfire had intensified. So, Lt. Col. Green asked a series of further questions designed to ascertain who was in the buildings:

> No one had any idea who was in the other three buildings. No one really had any idea who was even in the main building. They knew "bad guys" were in there, so they said, but they had no idea who else might be in there already when they got there. They hadn't been there long enough. And at the end of the day the ability to maintain truth meant a lot to me. It meant you don't prosecute them, and you have options to gather more information and satisfy your rules of engagement more robustly.

So, it once again got heated. Lt. Col. Green said he was not giving air support, to which they responded they would "get it themselves anyway." Lt. Col. Green explained that he is the one in charge of the requests for air support, and he would not honor their request because they had not met the task force requirements. This discussion was "very, very heated." At this point, the commanding officer comes to the tent and expels the (highly vocal) US commander, leaving Lt. Col. Green "feeling pretty good" because it meant he had the leadership's backing of his decision to be conservative because of his concerns about fratricide: "I've gone my whole career without fratricide, not even an accusation, so it was important for me to play it safe. Unless my guys were in danger, but I didn't feel that they were."

The situation came to a head now, however, because the US SOF got on the radio and was trying to obtain its own aircraft. At that time, one could go direct to those in control of air support (rather than through Lt. Col. Green) and request emergency close air support. In order to qualify for emergency air support, it was required to have "troops in contact" (meaning troops have come under fire or encountered insurgent forces) and for troops to declare they were in an "in extremis" situation (this concept is discussed in-depth in Chapter 8). So, the US forces used this workaround to secure air support. Now, it is important to state that the air support it received was actually an AC-130 gunship (a heavily armed fixed-wing aircraft). Gunships, however, are not *close* air support, meaning that they do not act under the same doctrine and, hence, do not need the same degree of communication (or confirmation from troops on the ground). Instead, the gunship has authority to fire missiles without approval from Lt. Col. Green. So, the US forces call "troops in contact" and request gunship support. The gunship arrives and destroys the target (the four compounds from which gunfire has been emanating).

Later that evening, both sets of forces return to base having been out in the field for 2 or 3 days. The German forces actually ended up coming into contact with the insurgent forces and having their vehicle "shot up and a soldier shot in the ankle." When the commander entered the tent, however, "he just looks ashfaced. Like he had gotten an ass-sandwich." So Lt. Col. Green pulled him aside and asked, "What's to be upset about? We had two teams in contact, two teams suffering injuries but everyone is coming home alive and the army was able to strike the target?" "Well," he said, "that is going to be really fine as long as we are sure what target we hit." Right then, Lt. Col. Green realized that "something was not right":

> So within about 30 minutes we find out that their [the US SOF] map study wasn't good. And this is one of the issues over there, the little compound you are looking at looks just like the one on the other side of the road. And they were about a kilometer out. They weren't even at their objective. The gunfire

was celebratory fire from President Karzai's friendly forces coming back from a mission, or back from a patrol. So, they were drinking and shooting in the air. And the call for fire had killed 26 Afghan troops that had openly been deploying with us. They had been linking up with us and doing missions with us. So now President Karzai had the ass, the friendly Afghan Forces had the ass and we're just waiting for it to get in the news.

This is an unfortunate case, but it reveals the importance of considering the ecological niche within which decisions are made. Quite rightly, psychological theories focus on the mind of the decision-maker. But military decisions involve multiple decision-makers and multiple individuals. Hence, attempting to understand the nuances of military decision-making (and especially military errors) requires that we consider the psychology of the whole team, not just the central individual.

In the previous chapters, we have examined how decisional conflict emerges when one must juggle equally adverse options. However, once the decision-maker has been able to choose, conflict can still emerge. For example, it can stem from conflicts between the decision-maker and other members of the decision-making system. These types of conflicts can be a source of stress for the decision-maker and can be equally detrimental to outcomes. Lt. Col. Green clearly struggled with conflict as it pertained to deciding whether to deploy air support and, if so, to which team. However, by relying on doctrine and policy (as discussed in Chapter 5), he was able to make an effective decision because neither group had reached the doctrinal threshold for action (namely they did not have PID). In addition to juggling cognitive conflict, Lt. Col. Green's decision also created significant emotional conflict between the commanding officers of both the German and US forces:

And me in the tent, I've got a German Representative and a [US Army] Sergeant Major getting really heated. German's yelling "Hey, this is fucking bullshit. Every time you guys do anything the Americans have priority. What's a coalition if it's always your teams who have priority? Just because it's your aircraft and your command and your leadership doesn't mean we are any less of a value." So, they are finally calling a spade a spade because they see this play out a lot and they see the Green Berets squeak a little louder and they tend to get more attention. They are openly aggravated, but what I thought was fairly professional. The Sergeant Major in the Army was extremely vocal, and he was a typical SOF Sergeant Major meaning he was loud and using some very creative language, re-inventing cuss words in new and exciting ways. He was extremely vocal and yelling. It took probably three or four times of me saying "Settle down, I need space. I'm on the radios and if I lose the radio conversations then I lose SA, and if I lose SA then we can't make good decisions." So it became a bit of an issue and I kept reiterating to the Sergeant Major that he had not met my criteria and I was not giving him air. And I told the Germans the same thing. . . . At some point, the Commander came in and told the Sergeant Major to "leave me alone and get the fuck out of the way so I could manage my job." So he [the Sergeant Major] was starting to get a little bit heated that I wasn't maintaining control of that element.

Military and critical incident decision-making hinge on the makeup of the team because difficult, multistep, strategic and tactical decisions are required. Military operations often involve a multitude of independent teams from diverse organizations operating both collectively and independently. From Lt. Col. Green's statements, it can clearly be seen that each stakeholder within the decision-making environment has different priorities. Lt. Col. Green prioritizes avoiding fratricide, whereas both German and US forces prioritize air support. The commanding officers prioritize their own forces getting support to ensure safety. These values obviously and violently clashed—creating significant pressure for Lt. Col. Green to navigate interpersonal frictions within the choppy waters of the decision itself. Furthermore, this value clash resulted in highly maladaptive behavior because the ground forces ignored doctrine and ordered lethal air support anyway.

As discussed in Chapter 5, least-worst decisions are driven by our values. Values are defined as stable, motivational constructs that guide perceptions, attitudes, and behaviors to achieve higher order goals (Schwartz, 1992). Acting against our values therefore creates internal dissonance because the action creates a disconnect between how we believe we *should* act and how we *do* act. Military decisions, as demonstrated previously, will rarely be the result of just one person's values, suggesting that multiple sources (people, agencies, and countries) of dissonance characterize difficult decisions. We need to closely examine clusters of values and beliefs and how they interact with one other (Liedtka, 1989) in military decisions.

VALUE CONGRUENCE

The previous case demonstrates a clear lack of "value congruence." Value congruence is the degree of agreement between the values of the person and those of the organization for which he or she works (Chatman, 1991). It is sometimes used to identify person–organization fit (Chatman, 1989; Kristof, 1996) and is strongly linked to job satisfaction and turnover because individuals whose values clash with the organization's values are far less satisfied (Rich, Lepine, & Crawford, 2010). As previously demonstrated, a lack of value congruence resulted in both interpersonal and organizational conflict. Accordingly, for military decisions, there are likely two forms of value congruence central to decision-making outcomes: value congruence within the decision-making team and value congruence between the individual and the organization. The case of Lt. Col. Green exemplifies value incongruence within the team.

CONFLICT BETWEEN INDIVIDUALS' VALUES

When examining value incongruence within decision-making teams, industrial and organizational psychologists have long debated the merits of group agreement. Within a team, conflict can stem from discrepancies and incompatible wishes or desires among the parties involved (Boulding, 1963). Intra-team conflict affects performance by producing tension, antagonism, and distracting team members from their task (Capt. Green's situation exemplifies all three). Through these mechanisms, conflict negatively affects team productivity (Saavedra, Earley, & Van Dyne, 1993; Wall & Nolan, 1986). Conversely, a lack of conflict and high within-group agreement (i.e., groupthink) can also be detrimental. Groupthink can generally occur when small groups seek uniformity of opinion rather than fully considering available alternatives (Janis, 1982). Groupthink is thought to have contributed to the mistaken assessment of weapons of mass destruction in Iraq, the Iran-Contra scandal (in which members of the US government were facilitating the sale of weapons to the embargoed Iranian government; Hart, 1990), and President George H. W. Bush's decision to remove Saddam Hussein's forces from Kuwait (Yetiv, 2004). Thus, some level of within-team conflict can improve performance by enhancing understanding of other peoples' viewpoints and creativity (Bourgeois,1985; Eisenhardt & Schoonhoven, 1990). In fact, diversity of personnel is an important component for organizational creativity (Hunter, Cushenbery, & Friedrich, 2012).

Diversity within a team, however, is a double-edged sword. It is worth specifying the nature of conflict that people experience when working as part of a team (Jehn, 1997). Overall, conflict types have been categorized as "emotional" and "task" conflict by researchers in this area (for other work on this, see Amason & Sapienza, 1997; Guetzkow & Gyr, 1954; Jehn, 1992, 1994, 1997; Pelled, 1996). Guetzkow and Gyr were the first to identify this distinction, and it has survived more than 60 years of scrutiny and empirical research. In addition to these core types of conflict, a third type of conflict has also been identified by some researchers (e.g., Jehn, 1997). Referred to as *process conflict*, this type of conflict centers on "conflict about how task accomplishment should proceed in the work unit, who's responsible for what, and how things should be delegated" and is separated from "'work differences' and 'petty b.s.'" (Jehn, 1997, p. 50; for a review of these three types of conflict, see de Wit, Greer, & Jehn, 2012). However, the majority of conflict research focuses on the two prominent sources of intra-team conflict: task and emotional (also referred to as "relationship conflict"). Relationship conflict stems from interpersonal incompatibilities between members of a team, and it is generally characterized as emotional because it stems from personal issues (e.g., dislike) among group members and feelings such as annoyance, frustration, and irritation (Jehn & Mannix, 2001). Task conflict stems from differences among opinions about how the task should be completed (Amason & Sapienza, 1997; Jehn, 1997; Jehn & Mannix, 2001).

Despite being an active hub of research within organizational psychology, psychologists have not yet examined the presence of these two types of conflict within military organizations and, crucially, the implications of this conflict on performance. Perhaps, given the importance of intersoldier bonds (as shown in Chapter 5), there might be an assumption that there cannot be inter-team conflict. Here, we try to show that this is not the case and that not only does relationship conflict exist within the military domain but also it can have significant impacts on how high-outcome decisions are made.

RELATIONSHIP CONFLICT IN IRAQ

Although there are often people in our working vicinity who we do not like, perhaps we underappreciate the impact this can have on performance. One Marine we interviewed provides a stark demonstration of the impact of relationship conflict in military decision-making. Corporal Park discussed a situation in which he (for a series of reasons associated with the experience of being in war) forgot to convey a piece of vital intelligence to some of the members of his platoon. Prior to our interview, Corp. Park shared his narrative account of the events with us, which is presented as follows:

> I was assigned to a Weapons Company Platoon as the Company Commander's radio operator prior to the war. The command element's responsibility was battalion fire support coordination (FSC) between the battalion line companies and the weapons company in conjunction with the artillery battery, the air wing, and naval gunfire support attached to our battalion.
>
> On the opening day of the ground war (this was my first experience in combat) my battalion was moving towards the second breach line located on the southern perimeter, which was being heavily contested by the ad-hoc Iraqi infantry in place. I was with my Commanding Officer (CO) in

an AAVC-7A1—command and control Assault Amphibious Vehicle—(AMTRACKS) which held the battalion FSC team (me, CO, and my platoon sergeant) the Air Liaison Officer, the Naval Gunfire Fire Liaison Officer, and the Artillery Liaison Officer, and the Battalion S-3 (Operations Officer) team. We were one of the lead AMTRACKS into the breach, as the S-3 Operations Officer was leading and directing the battalion at the time.

The 2nd breach line was being contested and we were under intermittent mortar, small arms, heavy machine gun fire from the various trenches and obstacles scattered in front of us. During this time, we had at least 2 F-18s on station to support the battalion—equipped with Mk. 82 GP bombs (or similar munitions). The S-3 requested that we mark the trench positions for the air element to utilize the aircraft munitions while they were on station (I want to say their attack window was approx. 15–20 minutes).

He requested our Heavy Mortar Platoon to drop white phosphorus rounds to mark the positions for an airstrike, then after this we could possibly follow up with a larger fire mission of regular high explosive mortar munitions. The mortar platoon was moved up close to our position—about 1000 m from the trench line and set up about 200 m to east of my position. We had established radio communications with both the mortars and the aircrafts at the time and coordinated the marking and initial air attack. Two passes from the two F-18s targeted the first series of trench lines. They dropped 4–6 rounds from what I could see from my spot on the roof next to the S-3 cupola of the AMTRACK—my CO requested that I watch to ensure the mortar rounds had made their mark correctly.

To the best of my ability, I could tell we had marked correctly and hit the intended target due to the spot of the bomb impacts in relation to our mortar rounds and the enemy muzzle flashes seen from the trench line. There was a pause—maybe 30 sec–1 min. However, the Iraqis resumed returning fire. Thus, the S3 requested the fire mission from the mortar platoon since we had what we believed were the right coordinates due to the presumed accuracy of the original phosphorus rounds marking. Listening to the radio traffic, it sounded as if the planes were now off station.

But unfortunately we could not reach the Fire Direction Center (FDC) of the mortar platoon—due to a faulty radio link or something of that nature. My CO told me I needed to go and inform the mortar platoon of the fire mission needed and ordered me to run to the platoon still located east of our position and pass the message. I was to inform the FDC chief of necessary sight/aiming corrections for the guns that he felt was necessary and wait for his order to fire—the coordinates were adjusted to better target the concentrated fire that continued to "harmlessly" drop in on our current position.

I climbed out of the AMTRACK—and began to run the 200 m to the mortar platoons position. During my travel to the platoon—along with the sporadic fire coming from the trench line, as well as my battalion's small arms fire, two aircraft (not sure if they were additional or the same) targeted the trench line (900–1000 m in front of me) again with what seemed to be cluster munitions and cannon fire this time. Being so close and unaware of the ingress the aircraft, the low flying aircraft and the explosions startled me and as a result I instinctively hit the ground. Additionally, as I was running, I had become tangled in antitank guided missile wire that was strung across the desert floor (it was either from the helicopter used in the opening month of air attacks on the positions prior to the ground war—or it had been possibly Saggers fired at us from surface mounts within the Iraqi positions). I know this because we had not fired any of the battalion tow or dragon weapons yet in the course of the operation.

Due to the events that were unfolding I forgot some of the message to pass to the mortar platoon. When I finally arrived, 2–3 minutes later, I was slightly shaken and I failed to pass the message on correctly. When I had reached the platoon—things were very chaotic and I could not locate the Lieutenant who was in charge. I passed the message on to one of the FDC members who had seen me running towards the location and met me to see what was going on. Due to my confusion and shaken state, I essentially got the message out but failed to pass on the desired corrections in their entirety—telling him to set up the barrage as is and wait for the order to fire.

As I ran back to the AMTRACK, I began to realize my mistake, but I chose not to go back and tell them. I was banking on the fact that the last pass of aircraft had done enough damage and that the fire mission would not be necessary. From what I can remember, as I ran back to the AMTRACK—I suddenly felt very exposed and almost oblivious to the enemy situation in front of me. I really just focused my energy on returning to the "safety" of the armored personnel carrier. In hindsight, I believe I was fearful that they would leave without me. Being alone and exposed between the mortar platoon and the AMTRACK was now very apparent to me—I also had left the AMTRACK without my rifle and was basically unarmed (minus my K-Bar knife and two grenades on my web-gear).

My quick reasoning told me that by going back to the mortar platoon that (1) I was risking my personal safety further and (2) I would possibly confuse the situation more, delaying the mission further. I felt that the first "spotting" mission was most likely accurate enough from what I had seen. So if needed, the mission I had passed on would be sufficient with the coordinates previously used. I was also hoping that by then the radio problem would be solved and thought that the fire mission could be double checked before being fired.

When I returned to the AMTRACK—from my perspective, the situation was still not resolved on the trench line—however there were other options being discussed than the mortar mission. Although this was still a possibility, it was not necessarily going to happen the way that we had originally planned and was being reconfigured accordingly. I felt that I had made the best decision not to confuse the chaotic situation anymore.

This is the most vivid memory I have stemming from my experience during the war. I have spent a lot of nights reevaluating my decision to do what I did and second guessing myself. Luckily, there was nothing to come of it.

There are many complex factors at play during Corp. Park's decision-making. As he highlights in his narrative, this was his first time in combat and he was, as he highlights, overwhelmed:

I was just overwhelmed. I had probably been up for 12 hours, with 2 hours' sleep, on top of two days of lack of sleep so my judgement probably wasn't as sharp as it could be. I think the realization that I was getting shot at was still haunting me; and not being able to pick up right away that I was getting shot at, and I had to look to someone else's reaction to realize we were being shot at. I am a trained Marine here and I'm in Disneyland. This is what I have been wanting to do my whole life was to be in a war. Literally since I was 5 that was my goal. So I was kinda in a dazed sense, a confused sense and then when the reality hit, and "Oh shit. This is war." So I kind of panicked, not "beserked"[1]

---

[1] Berserking is a physiological and psychological stress response to the experience of being in combat. As Farrell (2011) states, "In combat beserking is an emergency response to stress physiology. Overtaken by terror and exhaustion, Soldiers may charge into enemy fire, indiscriminately violent. They appear carried away or deranged. Fight overcomes Flight. It is plain. . . . The beserker's brain and body function are as distant from everyday function as his mental state is from everyday thought and feeling" (p. 9). Rather than a relevant psychological phenomenon, more recently, "beserking" has become a Hollywood style of combat depicted as "strategic fury" (for a Hollywood-ized example of this phenomena, see any Sylvester Stallone *Rambo* movie; for a discussion of this, see Farrell, 2000).

or anything, but mind numb.... The fact that I had trained and trained and trained and when the time came I was completely dazed to me felt like I was worthless ... I really think it was just "Holy shit, I am in the middle of a war in a really big area, and people are shooting at me."

And there were other factors that compounded upon this; tripping on the wire made him "feel like an idiot," leaving his gun still gives him concern today when he reflects back on the event. The "uncertainty of the situation, not wanting to let my Captain down, not wanting to let my fellow Marines down" all also played into his decision-making. But more important for our discussion is the relational conflict that interferes with his decision-making:

One of the FDC guys came running up to me, and he was a guy that grew up in my state with me. So, we were friendly towards one another, because we grew up near one another. We didn't know each other prior [to service in the Marines]. I mean we really didn't like each other, other than we were Marine buddies, there was just this kind of ... he was kind of a douchebag.

When asked about if there was anything else that could have changed his decision-making in that environment, Corp. Park responded that *"not running into that guy that I didn't like"* might have changed his decision entirely.

This case thus reflects what has been seen elsewhere in wider organizational behavior—that conflicts based on personality clashes and interpersonal dislikes are detrimental to group functioning (Amason, 1996; Jehn, 1994, 1995, 1997). Relationship conflict causes anxiety, which inhibits cognitive functioning (Roseman, Wiest, & Swartz, 1994) and distracts team members from the task at hand (Wilson, Butler, Cray, Hickson, & Mallory, 1986). In this case, although thinking that someone we work with is a "douchebag" may seem trivial, this simple friction had secondary effects on Corp. Park's decision-making. When reflecting on what might have changed the situation, he specifically points to his personal issues with this individual. Relationship conflict can also lead to sinister attributions about other group members' behavior (i.e., perceiving that they will, or are, acting with malevolent or hostile intent), which can lead to mutual hostility and conflict escalation (e.g., Baron, 1991). Thus, despite a lack of any real focus on the phenomena, it is clear that relationship conflicts can, and do, interfere with decisions made by military personnel.

Task conflict, on the other hand, is often associated with good decision-making. This difference stems from two interrelated effects (Amason, 1996; DeChurch & Marks, 2001; Jehn, 1995; Jehn & Mannix, 2001). The first benefit of task conflict is decision quality. Groups in which task conflict has occurred make better decisions because task conflict encourages debate, leading to better framing of the issue. Research has found that debate increases understanding at both the individual and the group level (Baron, 1991). The second benefit is that task conflict actually increases acceptance of the decision and leads to increased satisfaction within the group. This likely stems from the fact that task conflict affords individuals the opportunity to voice their perspective (Amason, 1996), and the feeling of "being heard" has long been associated with greater acceptance of group decisions, even when we disagree with those decisions (Simons & Peterson, 2000).

The effects of task, relationship, and process conflict on outcomes in decision-making are complex, and their impact is moderated by several factors, such as potential resolution (the degree to which a conflict can be solved), acceptability norms (group norms about conflict and communication), emotionality (the degree of negative emotion exhibited and felt during the conflict), and importance (the size and scope of the conflict; see Jehn, 1997). But what is more interesting is that studies often find that task and relationship conflict are often correlated (meaning that high task conflict is often associated with relationship conflict; see De Dreu & Weingart, 2003). This poses a problem for the view that task conflict is adaptive, because if, as task conflict increases, so too does relationship conflict (which is detrimental), how can task conflict have a positive effect on the situation? In the case of Lt. Col. Green presented at the start of this chapter, although the original conflict was task based (working out how we can best protect soldiers in the field while also maintaining the legality of our actions), it very quickly became a relationship conflict, meaning that differences in opinion were lost, overshadowed by interpersonal conflict.

Perhaps the answer to this question lies in *how* task and relationship conflict are interconnected. Several explanations for this correlation have been proposed (for a full discussion, see Simons & Peterson, 2000). The first view is that task conflict *causes* relationship conflict: Group members infer other members' intentions when they clash during a task, and if this leads to perceptions of a personal attack, or a hidden agenda, task conflict causes relationship conflict. Simons and Peterson referred to this as the hypothesis of misattribution and misinterpretation. A second explanation is that poor reactions to task conflict create relationship conflict. Specifically, if an individual uses harsh or offensive language to try to solve task conflict, he or she can trigger relationship conflict. Finally, it is suggested that relationship conflict may *cause* task conflict because an individual seeks to "sabotage" the influence of someone he or she does not like by manufacturing task conflict. Here, a more palatable "task conflict" is created by the saboteur to mask his or her (publicly frowned-upon) dislike of another team member.

To identity which of these explanations is most likely in real situations, Simons and Peterson (2000) examined the relationship between task and relationship trust in 70 of the top management teams in the United States. Specifically, they measured interpersonal trust and the use of aggressive conflict management tactics to gain insight into how task conflict might cause relationship conflict. Their results were strongly consistent with the hypothesis of misattribution and interpretation, in that "groups with low levels of intragroup trust displayed a much stronger positive association between task conflict and relationship conflict than did groups with high levels of intragroup trust" (p. 14). Ambiguous behavior (just like ambiguous visual perception) is often subject to top-down expectations; hence, when one person does not trust another, he or she is more likely to interpret ambiguous behaviors as sinister in the other person's intent. Simons and Peterson's data were therefore "consistent with the notion that intragroup trust is key to preventing task conflict from escalating into relationship conflict, presumably through an attribution mechanism." Later work has repeatedly supported this finding and the importance of trust within decision-making teams. For example, Peterson and Behfar (2003) examined the development of relationship and task conflict over time across 67 groups. Specifically, they examined the emergence of conflict after the teams had been given negative feedback on their early performance. Negative feedback can often create relationship conflict by creating animosity, antagonistic attributions, and hostility (Jehn & Mannix, 2001). Peterson and Behfar found that high trust buffered against task conflict turning into later relationship conflict, even in the face of negative feedback regarding early performance. In the absence of trust, negative feedback amplified early levels of relationship conflict (pre-feedback), and thus task and relationship conflict were positively correlated. What we learn from the business world is that the main mitigating factor in the emergence of detrimental relationship conflict in the presence of adaptive task conflict is trust.

TRUST IN MILITARY TEAMS

As with many psychological concepts, there is no universally accepted definition of trust. But, generally, trust is (Mayer, Davis, & Schoorman, 1995)

> the willingness of a party to be vulnerable to the actions of another party based on the expectation that the other party will perform a particular action important to the trustor irrespective of the ability to monitor or control that other party. (p. 712)

Trust therefore involves being in a state of confidence about the motivations and actions of another (Costa, Roe, & Taillieu, 2001), and often "trusting" puts the person in a state of risk, meaning that if the trustor is betrayed, he or she will suffer a cost

(Boon & Holmes, 1991). Trust, understandably, is foundational to military success. Soldiers "must be able to trust their colleagues, knowing how they will conduct themselves under fire" (King, 2006, p. 494). Similarly, US Army General Martin Dempsey has called trust the "single non-negotiable foundational value of our Army." It is no surprise, then, that research efforts, white papers, and military generals have all focused extensively on the issue of trust in the military (for review, see Stanton, 2011). The Army Leadership Doctrine, for example, implores leaders to "promote a culture and climate of trust" because a "failure . . . erodes unit cohesion and breaks the trust subordinates have for their leaders. . . . Broken trust often creates suspicion [and] doubt" (United States Army, 2012, ADRP 6-22, 6-7).

In military units, trust increases team cohesion and team morale (Cassel, 1993) and improves mission performance (Ivy, 1995). Shamir, Zakay, Breinin, and Popper (1998), for example, found that in the Israeli Defense Force, morale and unit cohesiveness were each positively correlated with the amount of trust that followers had in each other and their unit's leadership. It is no surprise, then, that trust was often discussed in our interviews. When referring to the calculus behind the decisions that they made, our interviewees often discussed the degrees to which they did (and occasionally did not) trust those around them. This modulated cognitive burden because without trust, constant vigilance is required to ensure people are not violating one's expectations (Luhmann, 1988). This benefit of trust frequently came up in our interviews, as servicemen and women noted that it afforded them a level of confidence in the decisions they were making and the ensuing plans they would execute. As one soldier recalls,

> When you had that much more autonomy it was really about trust. That's what it all boils down to; trust. Being able to trust subordinates to do the right thing and that they actually know what the hell they are doing . . . because if they don't know what the hell they are doing then they shouldn't be there.

Poor trust can have secondary negative effects on cognitive workload because individuals are unable to delegate tasks to "distrusted others." This is especially problematic in cognitively demanding settings (Alison et al., 2013). Lt. Col. Green's situation is an example of this because he had less trust in the US team and he had to burden himself with the issue of establishing PID:

> A key decision on my part was to engage with the PID issue. I very rarely, I can't even remember other instances where I was involved in the Positive Identification of targets when I was not on site. That is just not something I typically do. So that was a rarity. I think I just had a sense that "hey you guys

just got there" and I don't have a comfort zone with what you are saying. And that was rare. So as much as I argued that I can lose my credibility in the first 30 seconds, I think they lost theirs in my eyes and I was already questioning them and lacking the trust that I needed. So I think that was a two-way street. So the decision to get inside that decision loop of theirs I think was a very critical decision point for me, and very rare for me to do that. I'd like to think that I almost never do that.

Most often, our interviewees discussed having person-based trust, the type of trust that develops over time, based on direct experiences one has with another person (Rempel, Holmes, & Zanna, 1985). This form of trust is well outlined here by a former soldier:[2]

Just after being extended into OIF [Operation Iraqi Freedom], my cavalry scout platoon got attached to B/2-37 AR [B Company, 2nd Battalion, 37th Armor Regiment], a tank company, and we became "Team Battlecat." We began conducting operations against the Mahdi militia in Southern Iraq. The more we worked together, the more we trusted each other. When we were operating in the Kufah area, for example, my scouts discovered numerous mortar positions, which we handed off to the tanks to engage and destroy. The tankers' trust in our competence and our trust in them to support us grew with each patrol. Morale was always high when we got back from these types of missions, and there was a lot of high-fiving and sharing of stories from the engagements. The company commander was always checking on his Soldiers as well as mine after we had completed a mission. He always gave us valuable feedback, and we felt like we were accomplishing something important.

Here, trust was built with time and repeat engagements. This form of trust, which exists within a unit that has trained and operated together for a long period of time, is often highlighted in literature on military trust. Siebold (2007), for example, describes the role of trust and teamwork in that "the past twenty years, which does not dwell on intimate relations or masculine rituals but rather emphasizes interpersonal trust and teamwork built through many experiences including arduous training and drills"

---

[2] This quote is not from one of our interviewees but is instead recalled in a selection of shared stories of group cohesion, documented in the Company Command brief available at https://www.usma.edu/caldol/siteassets/armymagazine/docs/2013/cc_army_(apr2013)_cohesive_team.pdf.

(p. 291), and several of our own interviewees highlighted how trust in their team was developed through extensive training.

When examining decision-making, then, as mentioned in Chapter 3, military units do not always have long histories together and instead operate in the form of a "one-night stand" (they have a finite life span and are formed with a clear short-term goal or purpose; Ben-Shalom, Lehrer, & Ben-Ari, 2005). In such environments, soldiers often rely on "swift trust"—that is, trust developed quickly without direct experience with another person (Meyerson, Weick, & Kramer, 1996). Swift trust is often formed in ad hoc teams. This emerged in a study of the Israeli Defense Forces (Ben-Shalom et al., 2005):

> During the conflict, many of the regular frameworks of the military were broken up and new ones established. Such ad hoc frameworks—that seemed to work highly effectively—seem to contradict the image of "textbook units" marked by clear boundaries, continuity of membership over time, and strong internal cohesion. . . . These "instant units" were often composed of constantly changing constituent elements that came together for a mission and then dispersed upon its completion. (p. 64)

Thus, whereas some military bonds of trust are based on long-standing experiences working and training together, others are developed more fleetingly. But does this swift trust provide the same protective benefit? Meyerson et al. (1996) suggest that swift trust can be strong and resilient enough to survive the life of the temporary group because it is founded upon clear roles and duties. In temporary teams, relations are based more on roles than on individual qualities (because there is less face-to-face contact), meaning that swift trust is based more on professionalism than character: "This potentially 'cool' form of trust places less emphasis on feeling and commitment and more on action and cognition" (Huemer, von Krogh, & Roos, 2000, p. 135). And in most cases, swift trust developed between ad hoc military teams is sufficient for effective military operations. However, least-worst situations (from both a value and a psychological standpoint) provide a stern test to the trust that may exist within and between teams. How does swift trust stand up in least-worst situations? Is deep, interpersonal trust protective against decision inertia? Is swift trust based on roles or categories enough to facilitate least-worst decision-making? This issue has received little empirical attention. Although research has extensively studied trust (both long-term and swift in business organizations and the military), and conflict researchers have extensively studied the role of trust as a protective barrier from relationship conflict, such work is rarely applied to least-worst situations, especially ones that occur in physiologically demanding environments (e.g., special forces missions).

Whereas some aspects of trust (as discussed previously) are the result of strong interpersonal bonds, other bonds "have a finite life span, form around a shared and relatively clear goal or purpose, and their success depends on a tight and coordinated coupling of activity" (Meyerson et al., 1996, p. 167). In line with this view, research on responses to natural disasters has shown that swift trust can develop among complete strangers (Majchrzak, Jarvenpaa, & Hollingshead, 2007). In such situations, trust is often still granted on more superficial, surface-level factors such as shared categories (e.g., both being soldiers) or on the rank someone holds (one being a Major and another a Lieutenant). These forms of trust are referred to as "role-based" and "category-based" trust, and they can emerge without experience and contact with an individual but are granted based on assumptions about a category to which they belong (Kramer, 1999)—for example, trusting someone because you know he or she is a soldier or, specifically (as Col. Barker from Chapter 4 did), trusting someone because he or she is a member of Special Forces. Work by Ruark, Orvis, Horn, and Langkamer (2009) confirms this, in that they found that soldiers develop initial trust from surface-level factors such as rank and patches—things that *imply* individual qualities—whereas deeper trust only comes later through experience.

Thus, members of the military are able to use their experiences of training together with a unit to form deep bonds of trust. Although it is clear that trust is an important factor in building cohesion within a unit, based on what we have learned from studying decision-making in business organizations, trust also plays an important role in decision-making, especially in situations in which values can clash and conflict can emerge. Specifically, in such situations, trust within a team has a positive impact on decision-making because it reduces relationship conflict (Simons & Peterson, 2000) while enabling adaptive task conflict. Poor trust, on the other hand, results in maladaptive relationship conflict in the face of task conflict (and especially in response to negative feedback).

We can see these points in action when we look at the emergence of relationship conflict between Lt. Col. Green and the commanding officers involved in his decision. He highlighted that he had more trust in the German team and a greater comfort level. He was also well aware that the task conflict they were facing resulted in relationship conflict from the US Sergeant Major:

> It must have just been a comfort level with [the German] team. Which includes how often they checked in, and how detailed their information was. Because the Germans were very structured, they followed the routes that they were given, they communicated regularly. And their liaison officer was always with us and was always very patient.... So I think the comfort level with the Germans was much higher simply because they communicated better. So

I think communication, and rapport were the two big factors there. I had a regular relationship with the coalition teams, but that group [the US soldiers] was problematic at that point. . . . But in that period we had had more than one issue with that group, so I think there was a rapport issue. What went into that decision then was rapport, passion and emotion. Passion is good, but emotion is not. I want everyone to have passion, but emotions ran high and I think that weighed in, and when you have yelling and you have open emotions, and that guy (the US Sergeant Major), he isn't mad at the situation; he's mad at me. The German wasn't, he was frustrated at the situation. But the US Army guy was, he was mad at individuals and I think that breaks down a relationship, so I think that was another key factor in some of those decision points. Once it gets emotional, one of the golden rules as a leader is that you can't be the guy that throws a chair or blows his top. And I think he got there. He crossed those lines. That was one of the bigger flair-ups I have ever seen in the heat of an operation in terms of internal conflict. Typically, SOF was extremely calm and collected. You know, you can have dead guys and they will say "come on guys, let's keep our focus, we can grieve later." Remember "What's your objective." And then they talk through it. But at this point it was really heated.

It is important to state that cases such as this were rare; in the majority of cases we collected for this research, trust mediated relationship conflict, and most task conflicts were adaptive. But this case, which has been at the center of this chapter, highlights the importance of examining the effects that least-worst decisions can have on the responding team as a whole. Beyond merely placing the decision-maker in a state of conflict, they can often generate conflict between individuals, within and between teams, and between teams and the organizations for which they operate. In some (albeit rare) cases, these conflicts can result in maladaptive behavior that can have significant operational consequences (in this case, it resulted in the death of more than 25 members of the Afghan National Army at a time when such coalitions were both delicate and essential). To echo the adage we used in Chapter 2, in a situation in which the consequences of wrong decisions are so awesome, where a single bit of irrationality can set a whole train of traumatic events in motion, it is important that we truly appreciate the ways in which least-worst decisions can result in conflict (and hence potential irrationality) both within the individual and also within the environment within which they are making these decisions.

CONCLUSION

In Chapter 5, we highlighted how decision-making (especially least-worst decision-making) is driven by the values that someone holds, both sacred and secular. When people in a group have similar values, they are likely to prioritize problems, events, and goals in a similar way (Jehn, Chadwick, & Thatcher, 1997), leading to reduced conflict. However, when team members hold differing values, conflict can emerge (Jehn, 1994; Pelled, 1996). This thought often drives the erroneous view that homogeneous groups are the most effective because they will have greater value congruence (Adkins, Ravlin, & Meglino, 1996), but it was demonstrated in this chapter that some conflicts are good—especially those that allow for detailed discussion of the problem and the voicing of perspectives. This type of conflict is beneficial for decision-makers (and closely resembles the military practice of "red teaming"). However, task conflict can often result in relationship conflict because people either perceive task conflict as an attempt to discredit them or the task conflict is handled poorly (resulting in insults and relationship conflict). These interpersonal clashes can derail decision-making. In the example presented previously, relationship clashes led to a breakdown of protocol and communication and resulted in poor decision-making and the death of Afghan soldiers during a very sensitive period in the war in Afghanistan. But we also showed that trust can prevent adaptive task conflicts from morphing into maladaptive relationship conflicts: When interpersonal trust within a team is high, individuals are able to handle task conflict without becoming emotional. Here, we extended the role of trust in military teams beyond simply as a protective factor for those living and operating in extreme conditions but also as a protective factor for effective decision-making in the face of value clashes during high-stakes least-worst decision-making.

7

TEAM LEARNING

I am the leader of an orchestra. Here are the English bassos, here the American baritones, and there the French tenors. When I raise my baton, every man must play, or else he must not come to my concert.
—General Foch, General-in-Charge of the Western Front, World War I

The final phase of the SAFE-T model is team learning. It seeks to capture the process through which responding teams observe and learn from decisions that they are making and update their future decisions. In this chapter, we take a slightly different approach to the view of team learning and focus on the interoperability of teams within a system. As discussed in Chapter 6, least-worst decisions can create huge conflict between individuals. In this chapter, we further widen the lens to look at conflict that can occur between different teams that are all operating toward the same goal.

In many military and critical incident situations, teams operate as part of a coalition or collective of organizations. For example, in Afghanistan and Iraq, members of the military often operate alongside local army and police forces, multiple nongovernmental organizations, and other diverse military units with different capabilities and mission objectives. Interoperation adds complexity to the decision-making process because it brings together organizations with (potentially) disparate (and competing) values, priorities, and organizational practices. Such pressures are not accommodated by either doctrinal or normative models of decision-making or practically driven models based on recognition priming and experience. In this chapter, we explore the unique pressures of interagency military operations, highlighting both the lateral and the vertical conflict that can occur in such operations and the forms of uncertainty this can bring to the decision-making process.

INTEROPERABILITY AS "TEAM LEARNING"

In the SAFE-T model, "team learning" is an iterative process of learning throughout a major incident (van den Heuvel, Alison, & Crego, 2012). This stage has two components: (1) corrective, in which teams self-correct by learning from the mistakes they have made in earlier phases of the decision-making process (Salas, Rosen, Burke, Goodwin, & Fiore, 2006), and (2) monitoring, in which team learning is the feedback provided by other agencies on how groups are performing collectively and independently during major incident response (House, Power, & Alison, 2014). The corrective component reflects progress toward achieving specific goals over multiple decision cycles (in essence, it is an intradecision phase and reflects overall learning within a network). It views decision-making teams as "learning organizations" that are able to receive, collect, and adapt to the achievement of goals. Monitoring reflects the influence and input of other agencies during a critical incident. As such, it relates to interoperability in major incident decision-making and continual self-monitoring of the efficacy of joint performance. The military's ability to learn from itself (and from previous operations) is covered well elsewhere (e.g., Gerras, 2002; Posen, 1984; Rosen, 1991). Interoperability during missions, on the other hand, has received much less attention. Military operations involve multiple agencies operating in line with their own priorities and policies, and each operation is also guided by higher levels of command that also have priorities and values that may clash with those of the tactical or strategic commander on the ground. In this chapter, we outline some of the barriers to decision-making that emerge during interagency military decisions, as well as the barriers that occur when higher command priorities conflict with those of the officers at the strategic and tactical level.

INTEROPERABILITY AND EXOGENOUS UNCERTAINTY

Uncertainty (as discussed in Chapter 2) is a sense of doubt that affects one's ability to act (Lipshitz, 1997), and it is usually attributed to the situation, options, and outcomes that the decision-maker is dealing with—that is, the event itself. This is termed "endogenous" uncertainty (Lipshitz, Klein, Orasanu, & Salas, 2001). However, the event itself is not the only source of uncertainty. Alison et al. (2015) observed a live training event with senior police officers from the United Kingdom. During this event, the officers responded to a hostage negotiation simulation (in which civilians played the role of hostages and hostage-takers). The officers negotiated a series of demands (e.g., the hostage-taker asking for a vehicle to escape) while dealing with instances of harm to the hostages (e.g., one hostage had a thumb cut off). These are forms of "endogenous" uncertainty; they are associated with features of the incident. Throughout the scenario, the decision-makers were video-recorded, and the officers noted their actions in a "decision log," which asked them to "openly describe their situation assessments, recommendations for action, the actions they wished to implement, who they believed was responsible for executing these actions, and their rationale for taking these actions" (p. 299). Some of the officers were then interviewed in the weeks following the training event. These logs, interviews, and the video interactions were coded to identify sources of uncertainty experienced during the live scenario. In a situation designed to elicit confusion, the researchers were not surprised to find a high degree of uncertainty, but what was surprising was the overwhelming direction from which the uncertainty came. Specifically, whereas 43 of the coded discussions focused on endogenous uncertainties (a lack of situational awareness and uncertainty about the hostage-takers, their demands, intentions, etc.), 134 discussions focused on uncertainties about capabilities and trust levels present in the team responding to the incident. This form of uncertainty, independent of the events, we term "exogenous." We found that although both sources of uncertainty were present throughout the decision-making process, *endogenous* uncertainty most commonly emerged during the situational awareness phase of decision-making, whereas *exogenous* uncertainty was more commonly experienced in the plan formulation and plan execution phases of decision-making.

This research has two profound implications for our understanding of team learning and interoperability. First, uncertainty due to interoperation with other teams might swamp uncertainty arising from the situation itself, suggesting we should re-emphasize the importance of how multiple, diverse teams come together to operate as a cohesive unit. Second, because exogenous uncertainty was experienced throughout the decision-making process, we argue that "team learning" in fact occurs throughout the entire SAFE-T model of decision-making rather than being a discrete "phase" encountered after plan execution. Lessons from the hostage negotiation simulation (Alison et al., 2015) show that trust and role-confusion issues with interoperability derail both plan formulation and execution.

In the following example, we outline one case of interoperability between members of the International Security Assistance Force (ISAF) and members of the Afghan National Police (ANP). We introduce the reader to the specific challenges that working with coalition partners presents military decision-makers. This involves both specific issues, such as being able to trust those within other agencies (see discussions in Chapter 6), and more higher order issues that relate to cooperation between two large-scale agencies. Specifically, consider how the role of agencies outside of the decision-making team adds new elements to the equation and how the values and priorities of those higher up can diffuse down and affect a soldier's choice.

INTEROPERABILITY WITH THE AFGHAN NATIONAL POLICE

One of the unique pressures soldiers face arises from working with, and often building up, local military forces. Partnerships with indigenous forces have several benefits, including the ability to exploit local knowledge and force multiplication. However, they also have risks—seen most devastatingly in the emergence of "insider attacks" in Afghanistan in which members of the Afghan forces targeted members of ISAF. From a decision-making standpoint, such partnerships create new barriers to decision-making, including a lack of common language and varying levels of trust. Such partnerships generate even more barriers when members of the indigenous forces become the potential targets of military operations. One of our interviewees describes an operation he was involved in, in which a member of the ANP was linked to a local insurgent network.

In 2010, while stationed in Afghanistan, Maj. James received an intelligence report from one of his ANP sources that an Afghan police officer was tied to, or believed to be tied to, a certain improvised explosive device (IED)-producing cell. This was an internal mission with a turnaround (from receipt of mission to execution) of approximately 72 hours "because there was some expectation that anything beyond that would get back to [the target] no matter what, and that somebody would leak the information sometimes." The source himself had previously worked with Maj. James as an interpreter because he was too young to join the ANP, so in Maj. James' mind, he was a reliable source because he was a "guy that wants to actually do something for the right reasons as best he can within Afghanistan." Given his leadership of the battalion, "this operation was basically mine [to plan and execute]." So Maj. James sent out a platoon to analyze where the target was located,

> where he lived, family members and stuff like that. [We] actually did air reconnaissance as well because we have flights going out of there all the time. . . . We even went so far as to take four wheelers out to do reconnaissance over the area.
>
> So once we got a good idea that it was a pretty good target that we needed to take down we still wanted to be sure we took him alive if possible, but we needed to get in there to see what the situation was, what the risk was. The number of folks we identified on the compound ranged between 7–10. There were at least a couple females on the compound, and we knew that his brother lived out there as well, and we believed it was his father and uncle or something like that so there were at least 3 males. So we knew there was some potential for further risk there. We knew he was also a trained ANP officer so we knew he had weapons. So that's what we were going into in that respect. So once we'd identified what the target site was like, then we had to figure out "Okay how are we going to take this thing down?" That's where we come into the different dilemma.

To understand Maj. James' ensuing decision-making in full, it is important to consider the wider context. In 2009, there was a renewed emphasis from strategic command to involve Afghan local forces in missions. This, as Maj. James recalls, was "not as bad as much of the past year or two [2014–2016] where everything has to be Afghan-led, but the guidance was if you are partnering with these folks you will utilize these folks." This presented Lt. Col. James with four possible options: "We do this independently by ourselves and take them out, we do it with the Afghans jointly, we let Special Operations Forces [SOF] do it, or we have the ANP do it alone."

As Maj. James saw it, giving it to the SOF

> would have been the easiest way because it was out of my hands. I could have passed on the intelligence, my number two would have briefed them and I had a SOF detachment on my compound with me so it would have been the easiest thing.

The issue with this option, however, was the ramifications arising if the SOF raid went wrong

> But the process of figuring out how to deal with it again if something went bad with the raid, if the female had got killed or something like that because we literally spent months dealing with the previous incident before, and the amount of shuras[1] we had to have to deal with that and trying to calm everything down. I was very afraid of that happening again. So, although it was an easy decision that would meet the objective, ultimately we can't take the risk because the risk is too high, because it would increase the tensions too much if something went bad, even though I wanted to take that decision.

The next option was for Maj. James' forces to execute the mission themselves, which seemed like a simple decision as well, because "we could control everything and we had reconnaissance, we knew where everything was, could get the fire support from the Forward Operating Base back about 10 miles behind us and basically a conventional raid operation." It was a very "doable raid." However, although autonomy offered a tactical benefit, it potentially had a significant strategic cost:

> I live literally right behind the Afghan compound so I'm attached to the police compound so the variance of weight is on two factors. Firstly, if I execute

---

[1] A shura is a Middle Eastern dispute resolution custom, and it is usually an opportunity for members of a community (rural or urban) to resolve conflict and engage in joint decision-making. As the war in Afghanistan progressed, shuras were increasingly used to facilitate counter-insurgency and strong population–ISAF relations.

this will they maybe see me moving on an operation? And we did stuff all the time but this is obviously a different time of day, etc., I was a little concerned about that one, but I figured we could hide our movement in a way that they couldn't figure out where we were going. The second one was that I partnered directly with the zone commander, who's the 2 Star commander there, and then what's he going to bring up in terms of "you don't trust me," "you're going after one of my guys and you don't tell me about it, you don't have the respect to do this?" And this is about 10 months into our deployment, so we'd spent a lot of capital on building that relationship with the zone commander. So that was part of the concerns that I had if we had executed autonomously.

Ultimately, Maj. James decided to "put that one off to the side for now until we can see what everything else looks like." Before we continue, we pause to re-emphasize what is going on here. Maj. James is choosing against the most likely "best" tactical option because of the implications of these tactical actions for wider strategic concerns surrounding the overarching directives to work in partnership with the Afghan partners. This is important not just in terms of understanding how decisions are made during war but also to see the inadequacy of so many theories of decision-making that ignore these complexities and that change what we determine to be the "best" course of action.

The third option was to conduct a partnered raid with Maj. James' troops and the Afghans. This would meet the requirements and objectives of what General Petraeus wanted, in that it involved working jointly with the Afghans and would generate their capability and potential, but this could come at significant risk to US soldiers:

> I was concerned for my Soldiers' welfare. That was my biggest thing that came in for me, because there was not a lot of direct action tactical training and joint operations we'd done with the police. So, the majority of the stuff we had done with the police was training at their training base, at checkpoints with them, doing things of limited scope, and here we were talking about a night raid with them. I feared, not necessarily that there would be green on blue or anything like that which is always a possibility if you're in that environment, but more of a concern was safety from crossfire. Is there a way that we can set up the objective in the right way so we're not going to get in each other's way and be able to take the objective down? . . . Ultimately it came down to when I did my risk assessment, I'm like "I can't take that risk, it's a high risk that I can't mitigate enough to bring it down to a moderate." And obviously a moderate in combat is a high anywhere else but you can't go into

a high in combat because that's very high risk. So ultimately we decided that's going to be a no go on that.

The only remaining option was to allow the ANP to launch the mission alone. This option was Maj. James' "least-favored option" because it required putting all his trust in the Afghans:

> Even now having spent multiple years with Afghans and dealing with certain people in the same way, I think the only person I personally trust is my barber that I had there for the entire time that I was there. Oddly enough, he's about the only person. So I would never say I would trust anyone with my life. I would trust a lot of them sometimes to do the right thing, but ultimately it's going to have to be in their best interest. So that was the biggest thing, because we were always going to be secondary to their best interest and that's what it came down [to]. Not that they wanted to kill us necessarily, but "the Americans are leaving, so I'm going to do what's best for me." So that always went behind my decision process to deal with the Afghans.

This option required the ability to communicate the mission to the ANP in a way that would satisfy their needs and in which it would not be leaked too early, giving the target of the raid an opportunity to get away. Furthermore, this is already day 2, leaving only 40 hours to develop and train an element for some kind of raid (while still without telling them where or who the raid focused on) and still brief the ANP commander who would lead the raid.

Maj. James assembled his primary staff to red team through the process and discuss which option they would choose. Interestingly, however, Maj. James' plan (to give the mission to the ANP) did not "win":

> I did not win. I didn't even get my vote out initially but I mean my decision would have ultimately come out to be let the Afghans go by themselves. But this was not even selected by my staff. Which I don't blame them at all, they made the same decision using the same process as to why I wanted to choose some of the different decisions. Oddly enough the only person that decided to go with my choice, and his vote didn't really count for a whole lot, was my chaplain he said "Give it to the Afghans," which was interesting. But I ultimately overruled, not overruled, but I gave my guidance into it I said "Okay, I got your points, I got your decisions and recommendations on this, but we're going go with this [letting the ANP lead the mission] and here's why . . . ultimately what it came down to is we had been there for 10 months, we knew we would be turning over this mission to another battalion . . . and we knew we could not have a possible risk of having chaos for the next two to three months. It could go perfect, but it could go bad. And we knew that

we had to have some demonstration of their capabilities and trust with them for us to be successful down the road. So, this was really looking at a longer [more strategic] picture and further successes for the police versus the immediate objective piece.

So, Maj. James went to the Battalion Commander and got approval for ANP to lead the raid. They trained a small ANP unit up until 6 hours prior to the raid, at which point the ANP were given the full mission brief and they went and executed it:

> At 4 o'clock in the morning the ANP successfully got the objective, without killing anybody. The target did have caches of IED material and weapons and he did have ties to the enemy. So ultimately it was a success but it could have very easily gone the opposite way.

In the previous example, there are several clear exogenous sources of uncertainty that increased the level of difficulty for the decision-making process. Maj. James shows low levels of interorganizational trust. It is clear that the twin demons of competing strategic goals and low trust turned a simple decision (send in our own guys) into a potentially least-worst one. Although we have demonstrated how low interpersonal trust can cause maladaptive relationship conflict, here a lack of trust from an organizational standpoint (i.e., trusting in their ability to "get the job done") affected several aspects of the decision-making process. First, low levels of trust in the Special Forces (due to previous negative experiences) prevented Maj. James taking what he viewed as the "easiest option": giving the mission to SOF and having them conduct the raid. Similarly, Maj. James' low level of trust in the ANP also acted as a deterrent for using them either as partners or as the lead. However, once the decision had been made to use the ANP for the raid, low interorganizational trust still affected decision-making because Maj. James had to plan the mission without revealing the true target until a significantly late time (given that they were worried that the target would be informed and would flee). In addition, when the raid itself was occurring, Maj. James had made "preparations to support them if something went wrong. We had helicopters on standby, we had our vehicles ready and mounted if something went really bad. Communications all set up."

This case demonstrates what House, Power, and Alison (2014) refer to as "hurdles of interoperability." As they argue,

> The need for collaboration makes major incidents especially testing, since teams must coordinate as a collective whole by sharing cognitive aspects relating to agency-specific mental models (i.e. their knowledge and awareness

of what the situation presents) and areas of expertise to produce a coherent and collaborative response. (p. 321)

Despite a desire for interoperability, collaboration between multiple agencies is often ineffective. Power and Alison conducted a systematic review of the literature on both interoperability and decision-making, and they identified "trust" as a significant hurdle to interoperability. They note that "an interoperable command system should be based upon trust in other agency's job-related skills, personal motivations and adherence to in-group team values" (p. 326). In Afghanistan, there were low levels of trust in the agencies' job-related skills (ANP and, to a lesser degree, SOF), personal motivations (Maj. James argues they are "secondary to their best interest"), and adherence to their values. A lack of shared values is also a significant barrier to interoperability. Chapter 6 highlighted how decision-makers can experience conflict when their values clash with those of other team members; here, we see that value clashes might be even more likely when we take the interaction one level higher to teams of teams from different organizations interacting with one another.

Perhaps this point is worth emphasizing with a second example. Here, we reveal the conflict between tactical-level priorities and strategic goals and how this has significant implications for the choices that are made. In this case, the reader will, we hope, see how the course of action was chosen despite tactical success and to service a wider political/strategic goal.

BOMBS OVER BAGHDAD

By summer 2007, the Iraqi insurgency in Al Anbar province had started to wane. The surge, coupled with the Anbar awakening, had started to take effect and violence levels in the Haditha triad had significantly declined from 6 months prior. There were still insurgent attacks (infiltrations, suicide bombings, and sniper attacks), but these were not at the high levels that they were 6 months ago. During this time, Lt. Col. Chase of the US Marines was in charge of the Marine component in Haditha. One day during the summer, Lt. Col. Chase received reports at his command post that "one of the most notorious" al-Qa'ida leaders was in the area and that the Iraqi police had identified this guy and had chased him. However, he (along with several fellow insurgents) had escaped across the waterway and established himself on a small island in the Euphrates River. Lt. Col. Chase coordinated with a Naval squadron and the local Iraqi police to send some boats to pursue the objective. During the assault, the al-Qa'ida forces repelled the police attack, seriously wounding a couple of the police officers, causing them to retreat from the island and go back to the shore. At that point, the coalition and

Iraqi forces "fanned out across the shoreline and tried to contain the island so that the bad guys couldn't escape. And just, stayed in place there, observing the island and trying to keep it contained."

There were several possible options, including different kinds of attacks and approaches; however, it was not deemed worth risking more lives (coalition or Iraqi police). The Iraqi police had already been repelled once, and the insurgent forces clearly had automatic weapons, "so we just decided we were going to pummel the island with multiple 500 lb bombs." At that point in the war, it took a long time to generate the request (and then execute it); there were no air assets circling above, so they had to contact the regimental combat team (above battalion level), which then had to contact higher headquarters to get approval for air support.

The process of obtaining security air support took approximately 3 hours. The air support arrived and "whacked that island with enough bombs that everything would have been destroyed, any people on there would have been killed." Furthermore, the bombs "rocked the local neighborhood, broken windows, car alarms, things like that." However, given the delay in securing air support, the aircraft did not show up until after dark. As such, when the US Marines went to the island the following morning with Iraqi police to conduct battle damage assessment, they found that the insurgent forces had gotten away; they "likely jumped into the river and floated down stream right after dark."

In review, this seems like a simple case of waiting too long, thus providing a window of opportunity for the insurgents to escape. However, when Lt. Col. Chase delves deeper into his decision-making process, we can see that two unique forms of conflict emerged during the process. First, there is conflict between Lt. Col. Chase's chosen course of action and the course of action decided upon by his commander:

> My original reaction was the option of the Marines storming the island, or a combined effort. I think you are protecting the population more in the long run if you send in the Marines that way too. But the boss didn't see it that way . . . he was trying to contain the island as best we could from the shore, before he even had to make a decision, I mean I knew that was something we needed to do, so that was the initial push. But then his decision making, now I don't think there was very much debate in his mind from what I recall . . . not pursuing these guys to the fullest to ensure that there were no more casualties. So, it was very hands-on, he was kind of, he was fine, but he could micro-manage at times and he wanted me to feed him information,

he wanted information about what was going on, so he could inform higher headquarters. . . . Bringing in 500 lbs to me, you know, didn't make a lot of sense, and if I was the commander I don't think that that was the course I would have taken. But, you know, a counterfactual, it is hard to know.

Second, there was a conflict between the mission priorities of the US forces at this time and those of the Iraqi forces. In this case, the Iraqi forces' priority was to target the high-ranking member of al-Qa'ida, who is likely to pose a continued threat to them. On the other hand, this is not, in this situation, the priority of the US forces, whose mission is not to target members of local terrorist networks but instead to train, mentor, and develop a functioning and sustainable Iraqi military force. This disconnect reflects the difficult finesse required to balance counter-insurgency and counter-terrorism. Maj. James' conflict in Afghanistan (as outlined previously) similarly was that employing SOF (whose mission is often exclusively counter-terrorism) would have achieved the counter-terrorism goal but would have been in conflict with the wider counter-insurgency mission of developing a sustainable Afghan police force. This conflict is often further exacerbated because political considerations can provide even further top-down pressure for certain strategic objectives (e.g., minimizing American casualties). As Lt. Col Chase reflected,

> Well I mean, you know, public perception in the United States is extremely risk-averse, the war is unpopular, so there was pressure from all levels to keep coalition casualties low . . . we had al-Qa'ida on the run. With this tactical engagement that could have injured two or three Marines, Marines killed following these mid-level al-Qa'ida dudes, it was purely casualty aversion. . . . Yeah, having troops in the deck is always the better option for certainty in a mission like this, but our mission there was not to capture/kill al-Qa'ida, so tactically our mission was not to go in there and kill these guys.

Situations such as this can cause significant conflict because the decision-maker is forced to adhere to strategic command and implement a tactical directive that may not align with his or her own values and may not be the most "ideal" option. Such conflicts can be a source of significant stress for the soldier, especially because obeying orders is a prime military directive.

Interoperability between different levels of command therefore places the soldier in a precarious position because he or she may have a better understanding of the operating environment or the capabilities of the other teams/units at play but might be overruled due to strategic concerns. This is important because in many of our interviews, soldiers had to choose to challenge a direct order from a superior in order to prevent a military error. Sometimes, deciding which course of action to take is not a least-worst decision, but in fact, standing up to an authority figure is such a decision. Think for a moment back to the decision of Capt. Scott in Chapter 5.

The following case involves a military truck that was hit by a rocket-propelled grenade (RPG), causing a significant tactical issue (i.e., planning a mission to safely retrieve the asset), as well as a strategic issue (the potential for US military equipment to feature in enemy propaganda and the problems caused to members of the local population). The reader will see the interplay between the tactical commander's decisions and those being encouraged by his superiors.

"RUSHING TO FAILURE"

Given the threat of embedded IEDs in Iraq and Afghanistan, any land vehicles require significant reinforcements (often well over 10,000 pounds' worth of additions; see Roach, 2016) to mitigate the lethal effect of roadside bombs. This makes maneuvering such equipment both to and in theater a significant task. Lt. Col. Davids was in Afghanistan in 2013 when one such piece of equipment was hit with an RPG while being transported to base on the back of a truck:

> We had a piece of military equipment; you're transporting equipment back-and-forth in Afghanistan, probably 2013. . . . So it was riding on the back of a truck . . . and we had used a commercial hauler to do it because we were essentially bringing our kit into country and we didn't have our own trucks to do it. And the piece of military equipment which belonged to the US Army was hit with an RPG and the civilian truck driver decided that, because he had come under fire, that he didn't want to deal with it anymore and just offloaded his trailer and drove off and left our piece of equipment sitting in the middle of the road, on fire, burning. Tires were burning, engines on fire, and this just kind a left it there in the middle of the road.

Lt. Col. Davids first received a "SIGACT" (a report of significant activity) stating that a vehicle had been hit with an RPG in the local town and there was now a burning truck. After checking with several of the local forces (whose equipment was all accounted for), Lt. Col. Davids' unit realized it was theirs. Furthermore, this piece of equipment weighed 25–30 tons and was reinforced with another 15 tons of armor. To his superiors, Lt. Col. Davids had to answer, "How are you going to get this piece of equipment back?" Rushing into the middle of that area to get it was not a great option, but neither was leaving it where it was; it was clearly a disruption and a potentially valuable target for the insurgents. The vehicle was hit in a "very unfriendly" city, and launching a mission to retrieve it would have really caused "a spectacle":

> And so, we're looking at the options and they are bad and least-bad you know, do we leave it out there, and you know, get an IED planted on it? Or somebody can booby-trap it for you? And it was there, it was damaged, I don't have firefighting equipment, the tires are on fire and they are great big tires on this thing so it's probably going to burn for a day ... so that's what caused us pause ... [but] this is a major issue, it is blocking people out there, we have cut people off, so I think in the beginning, there was a drama, "We need to get someone out there right now, this is worth the risk of jumping into launching an operation" and it was exciting up front, and I think as we took a breath and said if we are going to go out there and do a mission let's do a quick mission analysis, and it's like—ok there is a lot of risk, it is just a piece of equipment and in a few hours the fire is going to go down. It'll be daylight we aren't going to go out today, let's not even think about it, and I took some frank pressure from those higher up in the organization to go get it but I said let's exercise some tactical patience and in the meantime let me talk to some units around and that's when we started to organize and arrange what equipment we have available to us.

Here we have a clear clash of priorities. Whereas his superiors wanted an immediate operation to retrieve the truck, Lt. Col. Davids wanted to exercise patience to properly plan the mission. The position of the burning truck, in the middle of a congested street in a busy local town, was clearly a hindrance. However, Lt. Col. Davids was able to get the street cleared with the help of a Polish patrol that pushed the equipment to the side of the road, at least opening the route: "This at least took care of the immediate emergency." This allowed Lt. Col. Davids to argue that he did not have to immediately put people's lives at risk: "I mean, it is going to be a risk to go get it but we can mitigate that to the biggest extent possible ... but the last thing we want to do is rush to failure." The next stage of the mission was action planning:

> So I think the next big decision was who is going to go do it. You know, who has got the capability to go do it, we weren't sure that we had the capability.... There wasn't sensitive equipment in there, so we weren't too worried about that, I think there was some GPS, some communications equipment but I think it was all burnt out. So we took a look at what we had available to go do it, looked at our own equipment, talked to our neighboring forces, specifically those that would provide air support, but they had other things that were higher on their priority list, but they said if we plan it out and got it scheduled, it was something that they would come do on a contingency basis.

Eventually, a second unit took responsibility and went to retrieve the equipment (without success). However, securing this was 'frictional" and relied on a combination of "FRIENDCON" and mutual support in other missions:

> It was frictional, I mean even us going and asking. They have their own priorities and missions and this is just a distraction for us. We almost had to go through another unit to even set up direct communications with them. And that's hard, just the communications piece. At one point I flew a guy up there just to have a face-to-face meeting. . . . So that was kind of the interaction with the Polish forces all of which had to be done through interpreters . . . and one thing we used during the interaction was our mutual support in other missions, you know and "don't forget that we are supporting you, we are supporting you down at FOB XX, and we are going to be working together for a while" . . . and you know that is how a lot of things are done in the military, through FRIENDCON I guess. So they come around, and were going to give it a go.

In the end, they were unsuccessful; after two attempts, they returned and said they were not going to try again—"It is too hot of an area" (during both attempts, they had come under significant amounts of enemy fire). Lt. Col. Davids said,

> They were just like, we can't do it, you can't run any of the controls, or do anything with it, it is just a big hunk of iron [and you should leave it there]. It wasn't kind of a frictional interaction—it was just kind of a sour talk to them about it.

However, they were starting to get a lot of complaints about this thing in the middle of the city, both from the local population and from other forces (that did not want to see US Army equipment burning on CNN or as part of terrorist propaganda). So, Lt. Col. Davids decided they were going to "figure out what our team looked like and put a team together." Furthermore, although the previous attempts had been unsuccessful, they had helped Lt. Col. Davids; they helped alleviate pressure he was receiving from higher levels of command who were requesting process updates and wondering "Why are you guys taking so long":

> I think at that point, because we had been working with those other units, each time someone would ask me, that's how the pressure starts "what are you doing about this," I had a plan. You know I always had a story to tell, "I'm working with the Poles, I'm working with Second Brigade." You know, this is the plan, they are going to go out tonight, "oh good," they are going to go out there tonight, and then when that fell apart, "well something happened out there, they didn't feel comfortable, they didn't feel comfortable with security," you know whatever the enemy contact was. They are going out there

again tonight, they are going to have more security, the Poles are going to help them, so then they . . . felt good about that. And I was getting a lot of questions, but as long as I had a narrative. . . . Then the second time, when they went out on, when they had such a hard time and they broke the pieces of equipment out there then I felt personally . . . kind of validated, my initial conversation with the boss that this is complex, it's hard, it's a bad area.

Second, even though the Polish forces were unsuccessful in their attempts, this, Lt. Col. Davids argued, gave his team a great learning experience:

Watching the unit go out there and try and do it twice, and they were supposed to be kind of the experts. My team sat there watching, stating "We could do that" and "We could have done that better," you know, these guys were taking notes, it was a learning process right, so we did that for two nights. . . . They had enemy contact the first night. The second night they broke two pieces of equipment and they weren't able to get it on the truck. We watched it fall off the truck and now it's in a really bad spot and I think I've got the guys and the equipment to do this, you know, we've been watching, we've been taking the lessons learned, and that's when we started coming up with a plan. I think I've got a plan to put it together, but I think we are going to need a lot more assets and this is going to take a couple of days for us to put this together.

Furthermore, because the mission was going to expose them for a large number of hours, they wanted to have a couple days to train for this complex event. However, the longer they waited, the more complicated it became because the chances that it would be exploited for enemy purposes (IEDs, snipers, using it as an ambush site, etc.) also increased. So, in the end, Lt. Col. Davids led a team of 80 to recover the equipment. They decided to cut the military equipment in half and articulate the individual pieces onto a recovery truck. They set up an inner and outer cordon for security (from potential sniper attacks) and used an explosive ordinance disposal team to sweep the equipment for potential IEDs. It took more than 3 hours to get there, more than 8 hours to cut the equipment in half and maneuver it onto the truck, and a further 4 hours to return it to base. The entire mission took more than 5 days to complete (from original SIGACT to recovery) and was, in Lt. Col. Davids' view, "a major military distraction, not a mission, a distraction, just a huge distraction."

Lt. Col. Davids' priorities throughout this mission were clear: Retrieve the military asset while ensuring the safety of his men. Strategic goals emphasized the negative political impacts of the military asset, increasing the importance of its recovery.

This created conflict because Lt. Col. Davids urged "tactical patience," whereas those in higher command pushed for a fast response to mitigate the situation. As Lt. Col. Davids states, in such situations it can take significant commitment to stick to a course of action in the face of pressure from those above you in the organizational hierarchy:

> [If I were more junior] I might have just listened to my boss, and gone when the boss told me to just do it. You know, we got to figure out and go do it. And I think that would have been an easy trap to fall into, a very easy trap to fall into, because it is hard to turn your boss down. Nobody wants to tell their boss "no," um, and I can't say that I was even comfortable telling him no. I just said listen, you know, let me do this, and let me follow up with you. And then I said, "I don't want to rush to failure. I don't want to let you down, I want to get it out of there too, but I don't want to put people at risk doing it." And that was a language he understood, and then we kind of stepped through it. But I can see, when your boss is mad, and I won't say he's yelling at you over the phone, but he's being pretty curt over the phone, uh, and maybe insinuating that you're not exercising enough initiative, it could have been a more reactive response.

Organizational hierarchies are, therefore, a double-edged sword because although they help ensure unity of effort and consistency within the many operating systems, a strict hierarchy and unified chain of command can restrict decision-making, leading to rigidity and risk-averse options due to a fear of "going outside the lines" or deviating from doctrine. For example, if an organizational hierarchy inhibits the degree of autonomy of an individual cell's confidence that its members can step outside of their operating procedures, decision-makers might fear breaching organizational taboos for fear of sanctions rather than being driven by actual goals. This is referred to as "threat rigidity," whereby decision-makers become risk averse in the face of a complex problem (Weisaeth, Knudsen, & Tonnessen, 2002). As we highlighted in our discussion on accountability in Chapter 3, during simulated counter-terrorism operations, police officers under high uncertainty often executed actions that would avoid future blame rather than actually move the situation toward a strategic or tactical goal.

INTEROPERABILITY AND INERTIA

Operating as part of multiple agencies slows decisions due to competing priorities and therefore increases the risk of decisions becoming derailed. Decision inertia is most likely to occur in complex and stressful situations, when information is of low quality and shared understanding is poor (Alison, Power, van den Heuvel, & Waring, 2015). Multi-agency responses exacerbate this because of the need to share cognition across geographically, and even culturally, diverse groups (Salas, Sims, & Burke, 2005). Thus, whereas group decision-making retards action (Greeley, 1995), decision-making in complex multi-agency situations more likely results in *failure* of action. This is because each agency (without the correct strategic direction or suitable time pressures) easily finds itself focusing on intra-agency issues rather than communicating and collaborating to solve the interagency problems so that the event itself can be addressed. An example of this phenomenon was recorded in Alison and colleagues' (2015) naturalistic observation of a 2-day training event involving a large-scale multi-agency (14 different agencies were represented) simulation-based exercise of an airplane crash in a major city. More than 190 senior officers from the 14 different agencies took part in the simulation. During the scenario, there were six critical incidents to deal with, including resource coordination between the multiple agencies, an emergency evacuation, and the handover of responsibility to local authorities. Throughout the scenario, the researchers repeatedly found that

> if critical issues were non-time-bounded, involved more than two agencies, and lacked clear strategic direction, then communications predominantly involved redundant intra-agency information seeking, as opposed to the useful interagency coordination of decisions and actions (as recommended by the SMEs). The decision making process was delayed due to decision inertia. (p. 314)

Another training exercise involved the police, fire, and ambulance services responding to a situation in which a train full of civilians was sprayed by a potential chemical weapon (later confirmed to be sarin gas). Despite the reports coming to the first responders quickly, and the majority of the fire service response team being ready to raid the train carriage, because of delays in the arrival of a delegation of police responders (who had the role of formally "arresting" the suspect), the raid on the train did not occur for 50 minutes, by which time—if this were a real event—all civilians would have been dead.

CONCLUSION

In this chapter, we took our discussion of conflict beyond the individual, beyond the team, to the role of conflict within and between responding organizations. We highlighted that the often heinous challenges of the event itself are not the only issues that decision-makers must deal with. They must deal with intra- and inter-team functioning, rely on one another, trust one another enough to communicate what needs to be communicated (and no more), enable each other to take on clear and noncompeting goals, and ensure that there is no repetition or redundancy in workload and efforts. We have also seen the strengths of goal-directed autonomy for boots on the ground versus the benefits of taking a "macro" level understanding of the battle space. There are no absolutes or certainties, and context, problem space, and type of incident will all be significant in indicating what method to use and when. Part of the problem resides in a philosophy of "one size fits all." In this chapter, we showed how strategic and tactical goals often compete; strategic goals often move tactical decisions away from the "ideal" course of action. We saw the conflict this can cause decision-makers as they must, sometimes, question higher level directives if they believe they have a better understanding of the situation. Once again, we see how sometimes least-worst decisions are the result of the responding *team* and not the decision itself.

In Chapter 8, we focus on the most difficult circumstance of all: how to make decisions "in extremis," or when one's life is on the line. There are unique challenges that lie within the *implementation* of a decision, and we provide the reader a sense of the process of making decisions *in a war zone*, not simply in war. So, for now, we leave behind the Byzantine negotiations of multi-agency, bureaucratic decision-making and focus our lens on the "mad minute"—those times in war when soldiers have no choice but to act or be acted upon.

8

LEAST-WORST DECISION-MAKING "IN EXTREMIS"

Everyone has a plan 'til they get punched in the mouth.
—Mike Tyson

The SAFE-T model has acted as a framework for our investigation thus far, but the model stops at the point at which the behaviors are taken to implement a decision. Where the SAFE-T model stops, execution begins. This chapter explores the unique psychological pressures faced "down range."

The term "in extremis" describes situations in which an actor is "at the point of death" (Kolditz, 2006, p. 657). Although many (but not all) of the cases presented thus far involve matters of life and death, very few involve decision-makers who are at the point of death (or at least perceive themselves to be). Studying decision-making in extremis is nearly impossible (Hannah, Campbell, & Matthews, 2010) due to the obvious ethical and pragmatic difficulties of developing data points when someone is in extremis. This is of significant concern because if, as the well-known military adage holds, "no plan survives contact with the enemy," then the successful realization of action plans made prior to the mission will need an extensive amount of decision-making in extremis. This means that what we know about military decisions leaves out the richness of evidence that could be gained from understanding how the extremis situation influences decisions. However, we can certainly infer that in extremis environments amplify the factors that ordinarily make military decisions so complex. Time pressure is increased; situational awareness is decreased; available assets to seek information may be minimal; and the saliency of the threat, risk, and accountability are greatly increased. Gaining a complete picture of least-worst decision-making in a military context requires us to examine such decisions in extremis.

It is widely known that stressful situations instigate the "fight-or-flight" response, the "immediate physiological reaction that occurs when danger or a threat to survival is perceived by an organism" (Milosevic & McCabe, 2015, p. 179). Bruce Siddle, who is on the board of Strategic Operations, provides a concise summary of the effects of the fight-or-flight response: "You become fast, strong, and dumb." This was recounted to Mary Roach (2016) as part of her recent study of some lesser known aspects of soldier science, such as sweat regulation, diarrhea prevention, and the potential future of genital transplantation for injured servicemen. Roach elaborates on the effects of the fight-or-flight response (2016):

> Our hardwired survival strategy evolved back when threats took the form of man-eating mammals, when hurling a rock superhumanly hard or climbing a tree superhumanly fast gave you the edge that, might keep you alive. A burst of adrenaline prompts a cortisol dump into the bloodstream. The cortisol sends the lungs into overdrive to bring in more oxygen, and the heart rate doubles or triples to deliver it more swiftly. Meanwhile the liver spews glucose, more fuel for the feats at hand. To get the goods where the body assumes they're needed, blood vessels in the large muscles of the arms and legs dilate, while vessels serving lower-priority organs (the gut, for example, and the skin) constrict. The prefrontal cortex, a major blood guzzler, also gets rationed. Good-bye, reasoning and analysis. See you later, fine motor skills. None of that mattered much to early man. You don't need to weight your options in the face of a snarling predator, and you don't have time.

Strategic decisions are almost never made immediately with one's life on the line. In this chapter, we therefore present a dynamic, slow-burn operation to show the complicated process of implementing a least-worst decision on the ground and, importantly, how soldiers react when a plan does not work. Finally, we examine the psychological factors associated with successfully (and unsuccessfully) making military decisions in extremis.

In previous chapters, we tried (where possible) to signpost readers to the relevance of the story and to provide commentary throughout the cases. Here, however, we choose not to. We want readers to consume this narrative and consider how they may react in a similar situation. During the course of our interviews, we heard about many very high-risk situations and many decisions that did not have positive outcomes. However, to us, few situations sum up the unique and phenomenally challenging psychological and physiological challenges that being "at war" entail. We hope the reader is able to get a sense of this from the following narrative.

"OFF TO A TERRIBLE AREA OF AFGHANISTAN"

Captain West deployed to Kandahar as a company commander in early March 2012. Capt. West had been on four prior deployments to Afghanistan—two deployments as part of an infantry unit (the main land combat force of the Army) and two deployments as part of a Rangers unit (an elite light infantry combat formation within US Army Special Operations Command [USASOC]). This was his first deployment as a company commander. His company, in contrast to its command, was short of both experience and manpower. Only a few company members had combat experience in Afghanistan, and none of his platoon leaders had been to Ranger school, a characteristic that Capt. West would have desired because they can "identify things, tactically, that guys that haven't gone, I think, don't." Overall, he viewed the platoon as "not in shape" and the general deployment as less than ideal:

> It was a short notice deployment, we were notified at the end of November that we would be deploying to Afghanistan. We were also short about 40 people, out of the 130 or so that I was supposed to deploy with. We don't have the physical skill set, we don't have the personnel skill set, we're doing a quick train up. We lose Thanksgiving week, we lose Christmas, and then we're off to a terrible area of Afghanistan.

Although Capt. West's area of operations (AO) was highly kinetic and enemy engagements were common, there was one specific village that was deemed a "hotbed" of enemy activity. It was also a known headquarters for the manufacturing of improvised explosive devices (IEDs) that were being used throughout the wider AO. Yet to date, the village had been subject to very few International Security Assistance Force (ISAF) operations. When Capt. West called the Commander (of that area), he found out that they were not even patrolling; they were just letting the Afghan Nation Army (ANA) "go out on their own." The insurgent group in this village was very well armed: "PKs, AK-47s, recoilless rifles. I mean you go in there and you find one that they've thrown [away] and you take it and a couple days later they just have another one." This village was therefore increasingly a concern for the safety of both ISAF forces and the civilian population in that area:

> Every time we went there; direct fire and contact. Nearly every time. We had, prior to [this operation] we had two or three amputees in the area from IEDs, couple of IED facilities that we had found, tons of pressure plates. . . . Very well done pressure plates, three to five pounds of homemade explosives

[HME], so you know just enough pounds to pop a leg . . . pop a leg off. And so we knew that this area was sort of an IED headquarters.

So, toward the end of the deployment, and after briefing the mission plan to the RC-South Brigadier General, Capt. West was given approval to launch an operation "to get in there and destabilize some of the things that [they] were doing."

THE PLAN

With support from the battalion, Capt. West's mission was to conduct clearing operations in two distinct clusters within this area. "Clearing operations" are a series of offensive operations (e.g., raids) aimed at separating insurgents from the local population. This specific operation involved three platoons—two platoons responsible individually for a single cluster of compounds and a third, smaller tank platoon providing logistical support to both. This 3-day operation was to occur in tandem with the ANA. On the first day, Capt. West deployed with Second Platoon, the larger of the two, and the operation went well, with "no issues really, [we] found a couple IED making facilities, [it was] really a great day." On the other side of the village, however, First Platoon got into a "pretty good" firefight, during which a squad leader (who Capt. West held in high esteem; he was an "absolute combat leader") was relieved from duty by the first sergeant:

> [The squad leader] was requesting additional ammo because they were just running out, and he thought that I was denying them ammo . . . so he said "I'm not moving," so he would not maneuver against the enemy and the first sergeant relieved him, so they lost that guy. A guy that First Platoon really needed, they lost him because the first sergeant pulled him right out of the fight and said, "If you are not going to lead your guys you're gone." So he went back to the base.

At the same time, this platoon was also down several other key personnel, including a platoon sergeant who had gone back to headquarters to prepare for their next deployment, as well as another member of the platoon who was evacuated when an IED blew up, damaging his shoulder. Second Platoon was now missing all of its more seasoned sergeants. In light of this, the following morning after the two platoons linked up and slept in an abandoned compound, Capt. West joined Second Platoon to conduct the clearance missions of the second cluster of compounds.

To reach the target compounds, the platoon had to navigate a large field of marijuana. Dismounted movement in Afghanistan is slow. In order to combat the ever-present threat of embedded IEDs, the US Army deployed a series of commercial off-the-shelf ground-penetrating radar technologies to detect low metal content IEDs (known as Minehounds or Vallons). These devices allow individuals to safely navigate IED-laden terrain in Afghanistan. However, it is widely known that successfully maneuvering these devices is "some science, but a lot of it is art." For example, Klein (2011) outlines his work with US Army Veteran Floyd "Rocky" Rockwell, who was able to maintain a 90% detection rate for IEDs, yet he had no overt awareness of the strategies he was using to do it. For Capt. West's operation (or any operation really), three Minehound devices would be required per platoon: two to sweep and a backup just in case. Second Platoon had one. This, coupled with the sporadic gunfire they were receiving and the discovery of a few pressure plates en route (which were quickly defused using an innovative lasso technique involving some detonator cord and C4), led to very slow progress through the field and toward the target compounds.

ENEMY CONTACT

During the grueling progress toward the target cluster of compounds, the platoon started to take very heavy fire from inside the village:

> There was a marijuana field, moving up to this cluster and we were in the middle of that . . . kind of [in the] open . . . when we start taking heavy fire, from PKM, AK47, rocket propelled grenades. One of the squad leaders was shot and then one of the ANA guys was shot as well, and so, you know we're out in the open, with little to no cover.

Capt. West and the platoon were able to seek cover behind a few small walls, while they radioed for Apache support as well as a medical evacuation. Capt. West cleared the Apaches to engage the insurgents in the village:

> I cleared them to start engaging into this village cluster. I mean, there was nothing else I could do, I knew what the collateral damage estimate was, I was just going to have to accept it, and so, we started engaging inside the village, it was to the point where it was almost like part of the village was on fire because we were shooting rockets into the village.

This provided relief for the platoon, allowing them to make forward progress to the medical helicopter. Thus (albeit after a series of failed landing attempts and one landing at the incorrect site), Capt. West was able to evacuate the casualties and move forward toward a point from which they could enter the village. Under pressure to get the platoon to cover, Capt. West took the Minehound (even though he had "no fucking idea how to use the Minehound") and cleared the building before the rest of the platoon moved into the first compound. Even when everyone was inside the first compound, they continued to take "pop shots, like someone would run around the corner and just spray with an AK47." After what transpired when approaching the village, "It was very clear that everyone in that platoon was very nervous." Although the platoon had temporary shelter in the compound, the individual operating the Minehound pointed out that they were getting "a lot of readings" (meaning that there was a high chance that IEDs were embedded in the surrounding area), limiting any, if not all, of their options to safely progress into the village and continue their mission:

> At this point I was still in clearance mode, I was like, we are going to finish this objective, because this was an entire objective we had to clear, so this was not, "Let's get out of here time" this was still "We have work to do." So we still had all of this to clear.

Options for safe maneuvering were low, so Capt. West maneuvered the platoon toward a small batch of pomegranate trees just outside the compound they had entered from the fields. This did little to increase their options. Minehound readings were still high, and the platoon leader was of the opinion that they should not go on:

> So I said, you know, so what do you [the platoon leader] think we should do? He said I don't know; we are just getting a lot of readings. And I thought to myself, we can't just do nothing, I mean this is the discussion I am having with them. Here we are, they know we're here, we've already had casualties, we are hearing traffic of them, they are maneuvering, so it's not that we can just say "I am getting a lot of readings," we have to make decisions.

The decision was made to continue with the clearance. However, there were few safe options to maneuver without significantly risking stepping on an embedded IED or walking directly into enemy fire, both of which could have disastrous outcomes for the platoon's already-dwindling manpower. The platoon now appeared to be in a complete state of inertia:

> This is a point that I identified, serious paralysis in this platoon, and so we had a discussion there, and uh, they just didn't want to go anywhere, so, could have gone back this way, so I was laying kind of laying here, talking with some of the guys about, you know, what our options were . . . all this time they said, it's not safe, it's not safe, it's not safe.

Captain West's solution to this paralysis was, in his words, "probably one of the scariest moments of my life."

Across from the pomegranate trees, on the other side of a small wall, was an open field leading to a small alleyway where Capt. West had "been in a couple of firefights" before. To get the platoon moving, Capt. West stood up, ran alone all the way across the field, and started providing cover to allow everyone else to get across. This decision put him at significant risk. During his sprint, he was both in the open and covering land that had not been swept for IEDs. However, it had the desired effect, and with him providing security, the rest of the platoon was able to cross the field and join him in the alleyway. This also allowed them to maintain their mission because they could now sweep down through the town and eventually reconvene at the exit point.

"BRING THE BIG GUN"

Although now able to complete the mission, the platoon was still facing threat from insurgent groups in the area, and the entire time they were in the village, the radiotelephone operator (RTO) was receiving "chatter" (intercepted insurgent-to-insurgent communications) about their movement. One of the predominant pieces of chatter that the RTO was getting was that the insurgent groups in the village were requesting for others to "bring the big gun." Capt. West could not be sure what this "big gun" was, but regardless, "It was something that we probably didn't want to face." At this point, the ANA partnered forces, and to a degree the entire platoon, decided that this was enough:

> So when we got in here, the ANA said "We're done. We're not doing anything else. We want to leave right now. It is dangerous here, we are not doing a single thing." Um. You know, I'm trying to talk to the leadership at this point, like, listen, this is why we are here, if we don't clear these compounds, let's say that there are 100 pressure plates, or 1000 lbs HME in one of these compounds and we don't get it, that, granted, what is the accessibility of HME to these guys? Maybe they have a lot of accessibility, but maybe they don't? Maybe they have to answer for the explosives that they lose. So, we're here, let's check it out. They wouldn't do it. No one else would do it.

Capt. West was now clearing the compounds alone, "just kicking doors and . . . clearing them. Maybe not the best decision, I don't know, but um, they weren't going to do it and they were very frank about that. So, I went in and cleared." At this point, the rest of the platoon was taking a knee or avoiding their roles in the clearance:

> You had people who knew that they were supposed to be clearing because we have graphics, we have all the, you know, this is our mission . . . who were like "let's get the fuck out of here," which is not, that's not, the primary thing, we want to complete the mission.

"LET'S EVACUATE"

After clearing a few more compounds single-handedly, Capt. West realized the platoon would not be able to complete the mission. Looking at their faces and gauging the energy level and level of commitment of the platoon, Capt. West realized he would not get any more out of them and decided to evacuate:

> I mean this is late in the day, and so another one of my thoughts when we are kind of here, we are thinking, the sun is going down, this isn't a scenario like in the Rangers where you have all sorts of assets stacked above you, unless you are in a fight you are not going to have committed resources to kind of watch you, so you know, sun is going down, the boys are tired, they are probably not responding that quickly, what other threats are in the village? We are already getting this ICOM [intercepted enemy communications] traffic—they are bringing in additional weapons.... This [was the] point where everyone was just kinda tapping-out and saying we don't want to go anymore.

Assessing their location, Capt. West identified that they could exit to the right of the compounds, jump across a creek, and filter down to the extraction point. So, after clearing the route, he walked down this path in front of the formation. Based on his expert knowledge from many patrols in areas like this one, he was "semi-confident about [a lack of embedded IEDs] because the ground was so hard and packed." When the ground is hard packed, it is more difficult for insurgent groups to embed IEDs and re-pack the ground in a way that does not show clear indicators such as distorted dirt (also known as "ground sign awareness"). Capt. West jumped between hard spots of ground before turning a corner to lead the platoon toward the creek.

As Capt. West turned the corner, two IEDs exploded—one right in front of him, one behind:

> I turned the corner, an IED went off right in front of me and an IED went off right behind me. The NCO [non-commissioned officer] walking behind me was thrown into a creek and then two other guys were wounded behind him on the path . . . then the medic . . . tried to advance and then, uh, one of his legs was blown off. At that point, I immediately thought that, well, I wondered if anyone was wounded, you know, that was the first thing, so the first thing, so, if I remember correctly, the first person I saw was the second or third guy, who was behind me and his face was just . . . just annihilated, I mean just blood everywhere, you know and then there was another guy who was just laying on a wall.... But there was no direct fire, it was just kinda silent.

The IEDs were "daisy chained" together, allowing simultaneous detonation once one was stepped on, and with the Minehound now snapped and in a tree, there was no safe way for any platoon member to move without risking the possibility of detonating more IEDs. Capt. West was effectively paralyzed. While other members of the platoon provided medical assistance to the soldier in the creek, Capt. West began radioing battalion headquarters and organized immediate MEDIVAC. While still separated from the rest of the platoon, Capt. West also started providing medical assistance to one of the injured soldiers until the UH60 arrived and evacuated four injured soldiers. After the successful evacuation, Capt. West was then able to link up with a scout platoon (a three- to five-man reconnaissance element), beefing up their security and allowing them to get the rest of the platoon to the exit site and back to base.

It is cases such as this that make military decisions so challenging—both to make and also, as psychologists, to understand. Here, tired, hot, and being shot at, Capt. West was required to engage in a level of cognitive rumination that would cause significant conflict. So, what happens to our theories of decision-making when faced with such acute physiological stressors? Next, we attempt to examine decision-making in such extreme environments. We outline the unique effects of them and delve deeper into the decision-making of Capt. West to examine how he, in comparison to the rest of his troops, was able to make decisions throughout this mission.

MAKING DECISIONS IN EXTREMIS

Capt. West and his platoon clearly had to navigate an increasingly complicated and risky operation that, from its onset, deviated from the original plan. They were undermanned, underequipped, and under constant enemy fire. The decision points Capt. West faced were almost exclusively least-worst. From a strategic perspective, abandoning the mission would mean potentially leaving a large cache of weapons or explosives intact and available for use against coalition forces in the near future. At the same time, continuing the mission would likely result in death or injury of members of the platoon. Thus, the calculus faced by Capt. West necessarily framed his options as negative on both sides. As the mission progressed, the tactical options available to Capt. West and the platoon were increasingly averse, with all possible courses of action likely to result in enemy contact or harm, as shown by the outcome of the eventual decision to evacuate the mission.

Capt. West's decision-making environment also shares several parallels with the decisions we have investigated throughout this book: There was a high degree of uncertainty, the situation was highly mutable, and Capt. West was accountable for both the mission and the safety of his men. At the same time, his decision-making environment posed a series of unique challenges due to the fact that these decisions had to be made in extremis. Thus, although the lessons learned from previous chapters apply in this situation, we must also consider additional variables in the equation.

Bearing the "Load"

Humans are only conscious of information that is currently available in working memory. Because working memory is capacity limited, it follows that our information-handling capacity might also be severely limited. In fact, working memory is widely accepted to be the bottleneck in human information processing performance. John Sweller's cognitive load theory, first proposed in the 1980s, makes use of the notion that the cognitive capacity of an individual—that is, the individual's ability to processes information—is finite (Sweller, 2004; Sweller, van Merriënboer, & Paas, 1998). This ceiling of cognitive capacity can be traced back to psychologist George Miller, who in 1956 published a famous experiment showing that, on average, the average human can store seven items (± two) in his or her working memory. Adding items beyond this capacity results in other items being lost from memory. Cognitive capacity is therefore in very short supply. Although a series of characters from popular fiction have shown the virtues of expanded working memory capacity (e.g., Michael Scofield in the television series *Prison Break*, whose low latent inhibition, a disorder linked to creativity and psychosis in which an individual is unable to screen out irrelevant information, coupled with his high IQ allowed him to solve complex puzzles, such as breaking his brother out of a high-security prison), the majority of us quickly bump up against the upper limits of our cognitive capacity when trying to process large amounts of information at once. Anyone who has tried to simultaneously pay attention to two conversations at once (as does any squad leader or platoon sergeant) will be intimately familiar with the limits that working memory places on human processing ability.

There are several types of cognitive load. First, load can be *intrinsic*. Intrinsic load refers to the number of items (or "elements") simultaneously being processed in our working memories. Simply stated, more elements mean more complexity, which means more cognitive load. Beyond this, interactions between elements multiply when we have more in working memory. Our prior experience with the items governs how they may interact in our memory. For example, a staff sergeant highly familiar with acronyms will be able to manipulate many such items while giving a situation report, whereas a private with little experience will quickly become overwhelmed. Although greater relationship between these elements, and greater expertise of the individual, can mitigate this (to a degree), a universal rule of thumb holds that the more elements to process, the greater strain we are placing on our cognitive resources (Sweller, 2004).

A second type of cognitive load is *extraneous*. Extraneous cognitive load refers to items that appear in working memory unintentionally; they are irrelevant and decrease the cognitive resources available to devote to intrinsic elements. In fact, research in the medical realm has shown that presenting patient information in a way that minimizes extraneous load (symbolically rather than textually) can result in faster reactions from physicians without a loss in accuracy (Workman, Lesser, & Kim, 2007). Accordingly, the US Army works continually with developers to find new ways of presenting battle-relevant information (on the ground or in the air) to help decision-makers integrate vast quantities of data. Current efforts spearheaded by one of us (Dr. Moran) at the US Army's Natick Soldier Research Development and Engineering Center (NSRDEC) aim to provide a standardized means to test future situational assessment technologies against their ability to provide soldiers with increased information while minimizing extra cognitive burden.

The final type of cognitive load is *germane*. Germane load (also referred to as "effective load") is the load incurred by the processes of elaboration and deeper processing afforded when intrinsic and extrinsic load have not saturated an individual's working memory capacity. Germane load in math might refer to the arithmetic operations we can perform mentally on simple information. Once the number of digits in our factors becomes too high, we cannot devote any more working memory resources to the germane task of forming our product. In this way, germane load facilitates effective processing of information, but only when there is enough space in working memory to devote to it (Anthony, 2008).

At war, all forms of cognitive load can be high. Consider, for example, the situation encountered by Lt. Col. Green, who had to decide whether to provide air support to US or German Special Forces teams. This decision required continual monitoring of new information coming in from each of the five ground teams he was monitoring (*intrinsic load*), as well as monitoring and responding to the heated interpersonal exchanges occurring within the command center (*extrinsic load*). To further drain his already finite processing resources, he had to continually generate a series of *possible* outcomes based on information he was receiving and act in anticipation of these predicted outcomes (*germane load*). In the case presented in this chapter, cognitive demands were equally high on Capt. West. At varying points, he was receiving and making sense of information from a range of sources, including readings from the Minehound, the movements and communications of the enemy, and communications with headquarters (specifically when organizing the medical evacuations). At the same time, he had to continually adapt his mission plan, respond to enemy fire, communicate and cooperate with the ANA (through a translator), and deal with casualties when they occurred. However, although the load of both was high, what differentiates Capt. West and Lt. Col. Green is that Capt. West had to deal with this load in extremis.

In extremis, a further factor vying for limited cognitive resources, is the *physical exertion* involved in operating as a soldier. Although we are all subject to some sort of physical exertion in the workplace, there are few environments that are as physically demanding as those faced by a dismounted (i.e., on-foot) soldier. So although others who operate in extreme environments, such as astronauts, climbers, or solo sailors, may share a series of psychological stressors (e.g., uncertainty, stress, and sleep deprivation; for an outline of the psychological dimensions of operating in "extreme" environments, see Barrett & Martin, 2014), very few other exertions also present the same physiological challenge.

Specifically, dismounted military personnel have to carry a significant external load. This is typically composed of protective equipment (i.e., body armor and helmet), combat equipment (i.e., weapons, ammunition, and power sources), and sustainment equipment (food and water). While bearing this physical load, they are also required to navigate diverse climates and terrains continuously for long periods. Ideally, a dismounted soldier should carry no more than 33 kg (~73 lbs, which is ~45% of the soldier's body weight; US Army, FM 21-18). However, although streamlining soldier load is often acknowledged, technological advances in firepower and protective equipment mean the average foot soldier is now carrying up to 60 kg (~130 lbs, close to the actual body weight of a soldier; Kobus, Brown, Wu, Robusto, & Bartlett, 2011).

It has long been known that excessive external load can negatively impact soldiers' physical capability and health (Bessen, Belcher, & Franklin, 1987). But what is increasingly becoming known is that this physiological burden also has significant cognitive consequences. Dr. Caroline Mahoney, cognitive scientist with the US Army NSRDEC, conducted a series of experimental studies with members of the US Army in which they were required to walk with and without obstacles and also with or without a 40-kg load. These soldiers were also required to complete a "vigilance task" by responding to visual and auditory cues presented on a monitor (Mahoney, Hirsch, Hasselquist, Lesher, & Lieberman, 2007). This research found that performance on the vigilance task was at its worst when participants had to negotiate obstacles and when participants were physically burdened with a heavy load. Follow-up work by Marianna Eddy of the Cognitive Science Team at NSRDEC examined how adding a standard load (40 kg) on a long march (2 hours) impacted soldier performance on a go/no-go task (Eddy et al. 2015). This task required soldiers to make a response whenever they heard the recorded sound of enemy weapons fire (AK-47) and to withhold their response when they heard the sound of friendly fire (M4). The study revealed that both adding a load and extended march time reduced cognitive performance in two ways: Soldiers were more likely to incorrectly indicate that they had heard enemy fire (raise a false alarm) during the second hour of the march, and when they were under physical load, the soldiers were less able to discriminate overall between AK-47 and M4 gunfire sounds both under load and during the long march. This diminishment of performance in a classic "cognitive control" task suggests that the capacity to use perceptual information to make appropriate decisions is reduced both when we have to be physically engaged for a long time and when we are under a physical load that is typical of those expected of soldiers. Similarly, when Marine infantrymen were required to walk for on a treadmill for 2 hours at 2 miles per hour (with a 5-minute break) carrying varying loads (no load, 44 kg, and 61 kg), they showed decreased accuracy and reaction times in a two-choice reaction test. Furthermore, this decrease was correlated with time on task; their effectiveness decreased the longer they walked (Kobus et al., 2011). This reinforces the previously discussed view that humans have limited processing capabilities that are further compromised during periods of physical exertion (although evidence suggests these abilities recover relatively quickly; Kobus et al., 2011).

Work by Sarah Smith and Graham Fordy (in press), psychologists at the Defense Science and Technology Laboratory at Porton Down (United Kingdom), examined the interplay between cognitive and physical load in dismounted soldiers. They showed that adding load to a soldier results in an increased demand for, and therefore reduces the availability of, cognitive resources for additional tasks. Thus, during highly physical tasks, the body and brain systems responsible for motor planning and control, rather than cognitive control, receive priority (as reflected in the quote from Roach [2016] at the start of this chapter). Neuroimaging work provides support for this hypothesis, showing that during exercise, activation increases in sensory and motor regions, whereas activation decreases in the prefrontal cortex (Diertrich, 2006). Furthermore, during this "arms race for energy," the tasks most vulnerable to being affected are those "that are novel and/or require focused cognitive control, tasks which require re-planning of activities and tasks that rely on real-time calculations and manipulation of information" (Smith & Fordy, in press). It is therefore the cognitive functions required to navigate least-worst decisions that are most vulnerable to being unavailable when under physical load.

In addition to load and the sheer length of typical soldier operations, there are other well-known barriers to cognitive ability that are amplified in extremis. For example, members of the Armed Forces often go days without sleep during deployment, often to the extreme detriment to their cognitive capability. Nathanial Fick (2006), a platoon leader in Afghanistan, recalls the experience of operating on little or no sleep when on deployment:

I had slept three hours in three days. "Gunny, I can't think straight. I need a couple hours in the bag," I said. At that point, sleep wasn't pleasant, just a mechanical necessity, like putting gas in a car. . . . It was the sleep of the damned. I floated in a netherworld of dreams, memories and sudden starts . . .

Christeson shook me awake. "It's been three hours, sir. The patrols on its way back in."

I sat up and rubbed my head, shaking gravel from my hair. "What patrol?"

"Team Three sir. They went to check out that tank." . . . Slowly, I understood. Some of my dreams had not been dreams. . . . I had given the order without even realizing it.

Sleep deprivation strongly impairs human functioning (Pilcher & Huffcutt, 1996): Those who operate for long periods of time without sleep are slower to react, make more errors, have trouble recalling stored information, and are less able to maintain attention (Barrett & Martin, 2014, p. 48). Tired people also show decreased creativity and flexibility as well as difficulties generating novel ideas (something that was definitely required in the case of Capt. West; see Harrison, & Horne, 2000). In fact, recent neuroimaging research, using functional magnetic resonance imaging (fMRI), found that when individuals were sleep deprived for 24 hours, there was reduced blood oxygenation (fMRI's indirect measure of neuronal activity) in areas of the frontal and parental lobe—areas critical for action planning and decision-making (Anderson & Cui, 2009). Chee and Choo (2004) found that sleep deprivation resulted in a "pattern of changes in activation and deactivation [that] bears similarity to that observed when healthy elderly adults perform similar tasks" (p. 4560).

Although military training often involves aspects of sleep deprivation, soldiers are still affected by poor sleep quality (Belenky, 1997). Perhaps a stark example is research conducted on sleep-deprived cadets of the Norwegian Military Academy who fired (what they thought was) live ammunition at human beings who had been substituted in for dummies in a live-fire exercise. Not only did 60% of those in the exercise shoot at a live target (moving versions of targets they expected to be inanimate cut-outs) but also, of the 40% who did notice that they were real people and not shoot, only one individual warned the others (Larsen, 2001). Larsen noted that "reduced cognitive ability due to severe stress and sleep deprivation may play an important part in decision making," at the same time acknowledging that because of their awareness of their own sleep deprivation, "it is also possible that some of the students chose to disregard their doubts as to whether or not to fire, believing they were hallucinating" (p. 97).

Soldiers, then, are under high cognitive and physical loads, performing tasks that are cognitively complex while under extreme sleep deprivation, during highly prolonged exercises, and in situations in which either their wrong decision or pure bad luck could get them killed. This perfect storm of factors, that singly and in concert guarantee a reduction in ability to perform, then combines with the expectation that soldiers should not only persevere but also excel. The military needs soldiers to be mentally adaptable in such situations (being able to adjust one's thinking in new situations to overcome obstacles; White et al., 2005, p. 14), but in reality, such adaptability is the cognitive function most likely to be diminished. For example, in 2005, the Army Research Institute conducted a study to investigate small unit leaders' adaptive thinking. However, contrary to being adaptive, the research identified that one of the most prevalent thinking and judgment problems was that small unit leaders did not "think on their feet" in novel situations. Instead, they often developed "tunnel vision," exclusively focusing on executing the mission one way while ignoring significant changes in the environment (White et al., 2005). When we consider the decision-making of those around Capt. West (although not Capt. West himself), we can also see actions reminiscent of an inability to engage in higher cognitive processing and cognitive inflexibility (the result being decision inertia):

> It was like "we are getting a lot of readings" and we got nowhere to go. This was the sense that I got at this point. So in [the lieutenant's] mind it was like we are getting readings here, we are getting readings here, we have readings here, we are getting shot at from here, someone came around the corner and I remember when this happened because it was like [gunfire noises] and the platoon leader he just screamed out "FUCK" . . . it seemed to him there was no solution to the problem, and maybe sometimes there's not. But at this point he presented it to me almost as if there is no solution. And so what I said to him was, what do you think we should do right now? And the answer was I don't know.

Navigating a derailing military operation requires a soldier to have access to higher cognitive functions, and our overview of current theories of attention and cognitive control shows how physical exertion, stress, tiredness, and anxiety all reduce the capacity necessary to engage these processes. We have also seen that at a neurobiological level, navigating least-worst situations depends on processes rooted in the frontal lobe, whereas being in extremis likely impairs frontal lobe function (although it must be acknowledged that the functional–anatomical characteristics of the frontal lobe are incredibly complex; Glasher et al., 2012). The frontal lobe integrates motivation, emotion, somatosensory information, and external sensory information to create unified, goal-directed action. It is thus the top of the hierarchy of neural structures devoted to the representation and execution of actions (current and future).

"Stress" in Extremis

The US Army (2009) field manual for combat and operational stress control highlights that

> no amount of training can totally prepare a Soldier for the realities of combat. Sometimes even the strongest Soldiers are affected so severely that they will need additional help. Combat and operational stress behavior experiences will impact every Soldier in some way. Just because [one] Soldier may not be affected by a specific event, it does not mean that every Soldier in the unit is handling the stress in the same way. (p. 1-2)

There exist wide-ranging psychological and physical reactions to being in an extremis environment. "Combat stress" and "operational stress" are the unfortunate reactions that may be experienced before, during, or after a soldier is exposed to combat (or combat-like situations; US Army, 2009). Combat stress involves experiencing personal injury, killing combatants, and witnessing the death or injury of a fellow soldier. Operational stress results instead from the prolonged exposure to operational environment, reduced quality of life, and separation from significant others and loved ones. Accordingly, the US Army aims to select individuals who are best able to handle the stress of combat, both physically and psychologically. Psychologist Salvatore Maddi (2007) has investigated the characteristics of individuals who are more capable of persisting through difficult challenges, developing the concept of psychological hardiness. Hardiness was found to mitigate combat stress in those who returned from the 1991 Gulf War, and the protective effect of hardiness increased with stress experienced on deployment (Bartone, 1999). Dr. Mike Matthews, head of Behavioral Sciences and Leadership at the United States Military Academy at West Point, further advocates hardiness as a desirable quality in soldiers, predicting that future selection tools will employ some measure of hardiness (Matthews, 2013). Despite this, it still remains almost impossible to predict how someone will really react when in extremis. This is for the simple fact that the data just do not exist from which we could make such predictions. When discussing his decision-making, Capt. West emphasizes the importance of considering the context of combat:

> This may seem a little funny, but you see people doing funny things after firefights and during firefights, like sometimes . . . you'll look over and see someone giggling and something, you just see odd things. Look over and you might see an Afghan National Army guy in the middle of a firefight sprinkling salt on a cucumber and taking a bite out of it. You know these are odd

things that, they just seemed out of place. Um. But at any rate I was doing an odd thing here in that I was having a discussion [about how to escape enemy fire] and I had a pomegranate open and I was eating out of the pomegranate. Maybe it was to try and calm down.

In Selye's (1936) first conceptualization of stress, he noted that the reaction was "independent of the nature of the damaging agent or the pharmacological type of the drug employed" (p. 32). Hence, it is the subjective feeling of stress, rather than the presence of a stressor, that dictates our reaction. It follows then that how different individuals respond to a situation depends at least as much on the individual as it on the stressor. Trait anxiety measures the degree to which a person responds fearfully to stressors (McNally, 1989), and high trait anxiety has been linked to poorer coping after extreme events such as hurricane Katrina (Weems et al., 2007), as well as an increased likelihood of suffering post-traumatic stress disorder as a result (Hensley & Varela, 2008). On the face of it, high trait anxiety would exacerbate issues for those operating in extremis. Accordingly, researchers have often sought to use trait anxiety as a predictor for reactions to, and performance in, military training. Survival, Evasion, Resistance, and Escape (SERE) training is a course aimed at ensuring an officer can survive on his or her own in any environment and under any conditions. During SERE training, students are required to (in an immersive training environment) evade enemy captors and, when inevitably captured, apply recently learned skills of resistance to indoctrination and captivity-related problems (Taylor et al., 2007). SERE training is deeply stressful and incorporates several aspects of an in extremis environment such as starvation.

The experience of anxiety (whether projected or concealed) culminates in the release of glucocorticoid hormones (Buckworth & Dishman, 2002). Cortisol is the primary glucocorticoid and mobilizes energy for action. Levels of cortisol in an individual's saliva are a reasonable measure of how anxious an individual is at that time. It would be viable to assume that those with high trait anxiety would elicit greater cortisol responses, showing that they are more stressed throughout training and, hence, likely to perform worse in extremis. This, however, is not the case. When trait anxiety and saliva cortisol levels of 26 active-duty Navy SEALs were measured (cortisol was measured both before SERE training and during the captivity phase), the researchers found that "with respect to the stress of the survival training, both groups demonstrated dramatic cortisol increases in response to SERE, but no group differences prevailed" (Taylor et al., 2008, p. 133). This research, coupled with several other trials of military personnel during immersive training (Rahe, Ryman, & Beirsner, 1976; Rose et al., 1969; Rubin, Miller,

Arthur, & Clark, 1970; Vaernes, Ursin, Darragh, & Lambe, 1982), supports that psychological dispositions alone are not predictive of stress response during military training.

The alternative explanation for individual differences in responses to the stress of combat is that they are process-driven. The process-driven model of stress response was developed by Lazarus and colleagues and involves two processes: the cognitive appraisal of the situation and coping (Coyne & Lazarus, 1980; Lazarus, 1981). Cognitive appraisal is the process through which an individual evaluates how a situation affects his or her well-being and involves two processes: a primary and a secondary appraisal. Primary cognitive appraisal assesses the importance of the stressor and whether the stressor is a threat (something that will cause harm), a challenge (a positive appraisal and something that can be overcome), or a harm-loss (something for which the damage has already been experienced). The secondary appraisal then decides what, if anything, can be done to overcome or prevent harm (Folkman, Lazarus, Dunkel-Schetter, DeLongis, & Gruen, 1986). These primary and secondary appraisals converge in a decision as to whether this person–environment "transaction" is threatening or whether it is a challenge that can be overcome. *Coping* is a person's cognitive and behavioral efforts to manage those aspects of an environment that are deemed as taxing (Lazarus & Folkman, 1984). When considering the platoon operating around Capt. West, these differences as well as (perhaps) individual differences in anxiety levels cause clear differences in their appraisal of the situation. Capt. West, while clearly aware of the dangers surrounding him, continually searched for ways to act and continue the mission (appraising the situation as a *challenge*), whereas those around him appeared to have adopted an appraisal focused on the threatening and harmful nature of the environment, highlighting instead the high readings that they were getting from the Minehound and the lack of safe options.

When trying to explain these differences, one important source of difference may have been that Capt. West had previously been in a very serious firefight in this town and had already found his way out from there, likely facilitating his mindset that this situation was a challenge to the mission, but one that could be overcome:

> Well I brought a team in here once and we got in a big firefight ... so I was here with four guys in this village sitting by this well, alone, so five Americans, well four Americans and an interpreter and we're hearing traffic like "yeah we see him," "we've got plenty of guys," I am thinking to myself we are getting ready to get killed, tortured and whatever. ... But we fought our way out is what happened.

Capt. West had experience with this village; he knew the village and had been in firefights in the village. But outside of this specific knowledge, Capt. West was also a very experienced soldier with four previous combat tours in Afghanistan, two with the Rangers. Ranger training is a multistressor environment that involves high energy expenditure, sleep deprivation, extreme weather conditions (hot and cold), anxiety, and physiological stressors such as infection. Soldiers on this course are evaluated for their ability to set aside basic needs in order to lead their patrol and accomplish the mission (Moore et al., 1992). Those who complete ranger training often suffer extreme weight loss (>13% of their overall body weight) and are more prone to illness (Pleban, Valentine, Penetar, Redmond, & Belenky, 1990). None of this platoon had been through Ranger School, and the majority of soldiers were on their first deployment. So, although this was obviously not anyone's first combat engagement (the company itself had more than 200 direct fire engagements during deployment), the platoon was still highly inexperienced compared to Capt. West. As Capt. West noticed, the platoon was very nervous. This is not to say that Capt. West was not nervous, but it is clear through his description (and, to a degree, his actions) that he appraised the situation differently and was able to focus those nerves toward achieving a goal. In fact, focusing on the negative aspects of a situation is closely linked with inertia. Nicola Power's (2016) work (discussed in detail in Chapter 5) suggests that inertia is linked with the behavioral inhibition system, which is involved in avoiding negative stimuli (rather than approaching it; Gray, 1990). It is viable here that by appraising the environment as threatening and harmful, rather than a challenge, others in the platoon were less able to engage in approach behaviors associated with problem-solving, finding themselves searching for a safe way out in a situation in which there really was not one. This would lead to inertia, as soldiers' continually attempt to converge on a course of action that is not realistic.

The Return of Recognition-Primed Decision-Making

After deciding that this was a challenging situation, and one that could be overcome, Capt. West still had to choose and act. Here, his experience in combat may have been a critical asset in allowing him to develop new courses of action. Let us consider one of the turning points in this operation: Capt. West's decision to sprint across a field (irrespective of the significant risk and potential placement of IEDs):

> I identified serious paralysis in this platoon, and so we had a discussion there, and uh, they just didn't want to go anywhere, so, could have gone back this way, so I was laying kind of laying here, talking with some of the guys about, you know, what our options were, so what I decided to do was to get them

moving, and the only way that I knew to do this was, um, to get them moving in a direction that, that they wouldn't have to go across to then give me a formation that they could say, well that's not safe. So all that they have said is, it's not safe, it's not safe, it's not safe. So what I did was I stood up and I ran.

When deciding to launch this course of action, Capt. West developed several hypotheses about the likely success of this plan. He knew that the area had not undergone any sweep for IEDs; however, because the terrain was an open field rather than a path, Capt. West believed there was a lower chance of embedded IEDs. At the same time, he also realized that moving the entire platoon (without anyone providing cover) would provide an easier target for insurgents rather than him sprinting alone (albeit slowly). More important, he thought the element of surprise would mean he was less likely to be targeted because platoons did not usually do that. In Capt. West's words, only "*idiots* do that." This highlights Capt. West's knowledge of the psychology of game theory, in which he reasoned (correctly, in this case) that because the enemy would not expect someone to be so foolish, they would not be vigilant toward this course of action. Such reasoning is not simple: It requires cognitive representation of typical standards of behavior, of one's own position, and of the enemy's likely mental representation of the situation.

Capt. West's previous experience was critical not only in allowing him to evaluate his new course of action but also in its original inception. When discussing how he came up with this new plan, Capt. West cited a previous enemy engagement in which a solo run in the open successfully alleviated the pressure on the rest of the force:

> We had a couple of combat engineers . . . they blew an IED . . . at that moment . . . a couple of single shots, the dog handler was shot, my platoon leader was shot, we scrambled back over the wall. The dog at this point is going nuts, biting everyone, the dog handler dies, but we were pinned down by machine gun fire and recoilless rifle fire. So we're pinned down behind this wall and . . . they were shooting from this wood line, so we are kind of behind this wall and then . . . probably 50 yards away was this little, I don't even know what it was, like a pile of dirt, there. . . . [So] I ran out this way and it took pressure off of these guys, again, not cleared, not comfortable receiving lots of machine gun fire at this position, but it takes pressure off and then it allows the medical helicopter to come . . . [and] it allows us to take pressure off them and apply pressure to them and it gives me a little bit better vantage point, takes me out of this situation and allows me to start calling in airstrikes deep into the cemetery and into the wood line. And it kills some guys. So this was probably a scenario that I drew off of.

You may already have identified the process occurring here. Capt. West engaged in recognition-primed decision making (RPD; see Chapter 2). He used his experience to generate a workable option, and once he had mentally evaluated the workability of this option (drawing on his past experiences to develop a hypothetical mental model of the outcome), he acted. In the same way, those in the platoon who perhaps did not have the same combat experiences as Capt. West were less able to develop parallels between the situation they were currently facing and those they had faced previously, rendering them unable to develop new courses of action. This presupposition has interesting implications. In Chapter 2, we outlined the extensive work of Gary Klein and colleagues and the process through which Klein proposes we navigate least-worst decisions. According to Klein, when faced with a least-worst decision, an individual is in the "zone of indifference," a point at which no option is beneficial and the decision-maker "might as well flip a coin." When faced with such decisions, an individual should not seek an "ideal" outcome but instead "satisfice," in that the individual should pick the first option that will get the job done. RPD may clearly be the avenue through which Capt. West's expertise and experiences allowed him to act. But given everything outlined previously, a lack of satisficing may not be able to fully explain the indecision experienced by the rest of the platoon because, we argue, they were not afforded the cognitive resources required to develop and (mentally) test courses of action to determine if they satisficed. In his own words, the platoon leader just "didn't know" what to do.

Beyond the tacit recognition that individuals who make least-worst decisions are likely to be under significant cognitive load, we have done little to delve deeper into the specific effect of this load on how they navigate least-worst decisions. Previously, we showed the cognitive drain, both physiologically and psychologically, of operating in extremis, but perhaps we are yet to truly appreciate the devastating effect that an extremis environment has on the ability to make least-worst decisions. In this case, the platoon leader did not avoid the decision (he could not because the platoon was in the open and under attack), he had not decided on a course of action yet failed to actualize it, nor was he currently evaluating a series of options and struggling to pick the best; he arguably did not have any options. This is important because in the absence of a similar experience, generating options to evaluate is, in itself, a cognitive burden; a degree of cognitive space is required in order to develop the possible courses of action to be evaluated. In this vein, without the available cognitive resources to develop new plans, individuals may find themselves in a situation of helplessness, in which although they know they need to make a decision, they are unable to engage in the cognitive process required to make one. In such extreme situations, inexperienced individuals may be more *decision incapacitated* than they are *inert*.

CONCLUSION

Throughout this book, we have discussed the hard decisions that Armed Forces often face. At the same time, we acknowledge that they are required to operate in an incredibly extreme environment. For those operating "outside the wire," this extreme environment is amplified due to the nature of the threat (i.e., an insurgent group that operates from within the local population) and the physiological demands placed upon soldiers. There are several theories of how individuals make decisions in such extreme environments, but we are rarely afforded the opportunity to explore individuals' decision-making in situ (instead relying on analysis of decision-making in simulations or during training). Academic research on the processes of decision-making in extremis will continue to be a challenge, and although this chapter presented merely one such case, it provides an important test of how well our theories apply to such situations. At the same time, this case is also an important addition to this book because it shows the unique stressors present in these environments and how these can confound an individual's ability to make *any* decision (let alone a least-worst decision). This is especially important because the military process often involves looking back and reviewing decisions made within an operation (especially those in which errors were made), but these reviews rarely consider the flaws inherent in the human (also referred to as the "human in the loop"). Finally, these considerations have wider implications for our expectations for performance in the field. Armed Forces need adaptive and flexible decision-makers, especially in hybrid conflicts such as those in Afghanistan and Iraq. In fact, the current Army Human Dimension Strategy (US Army Combined Arms Center, 2015) states unequivocally that the US Army must prepare leaders to thrive under chaos and ambiguity, in such hybrid conflicts just mentioned. Although this may be a reachable aspiration at the strategic and operations levels of command, at the tactical level, it is important that we understand how these extreme environments have the capacity to derail decision-making, rendering soldiers non-adaptive, inert, or even incapacitated. It is clear that research is required to understand human responses in extremis and to characterize the individual differences (beyond experience and RPD) that predict who will become incapacitated by such decisions and who will thrive.

9

THOUGHTS THAT HAUNT

Thy spirit within thee hath been so at war,
And thus hath so bestirr'd thee in thy sleep,
That beads of sweat have stood upon thy brow
Like bubbles in a late-disturbed stream.
—Henry IV, Part 1

Our story so far has taken us from the anodyne world of lab-based, sit-down decision-making tasks through the impressively increased realism of the naturalistic decision-making movement and, finally, to the field, where we have observed the inner workings of soldiers engaged in the most realistic—and critical—decisions of all. We turn our attention now to what happens *after* soldiers must engage in least-worst first decision-making and to the sequelae often observed after soldiers return home from theater—post-traumatic stress disorder (PTSD) and its effects. Three important facts stand out in the modern discussion of PTSD: (1) Many soldiers who engage in combat for their country return home with some degree of PTSD, (2) not everyone who is deployed will get PTSD, and (3) the protective factors that aid those with combat experience who do not suffer PTSD are still unknown. To complete the circle, then, we examine the potential association between making least-worst decisions and the development of PTSD post-combat.

Perhaps it is first important to clarify what we are *not* saying in this chapter. We are not saying that making a least-worst decision *will* result in PTSD. Nor are we saying that the people we interviewed, who have made least-worst decisions, *have* PTSD. However, we are saying that we believe that making least-worst decisions—decisions that are inherently negative and involve soldiers settling on a course of action that results in a bad outcome—leads to rumination; the consistent opportunity to wonder "what if?" once the decision has been made. Indeed, it would seem unusual for soldiers who have faced such decisions *not* to ruminate on how things could have turned out differently. It is this rumination, post event, about high-consequence decisions that may be one (of many) important factors in the onset of PTSD post-deployment.

In Chapter 4, we outlined the decision faced by Maj. R. Webb regarding whether to take offensive action at an oncoming SUV. When reflecting on his decision not to shoot a possible insurgent in a vehicle moving at high speed during the walk-in mission, Maj. Webb highlighted his tendency to still ruminate about that course of action several years after it happened:

> In retrospect I don't know if what I did was the right decision or not. He looked threatening and could have hurt, you know, uh, my friend, other Americans. And so, from the one side of me I still regret not seeing him and firing. On the other side of it, he was an innocent guy, and as far as I know he didn't intend any harm to us. But, I, uh, that was a tough thing for me . . . if I fire I kill an innocent person, if I don't fire I risk my friend's life. So, they talk about the training taking over at that point, and at this point I don't know, my training has taught me to do things like building clears, to defend myself in firefights, I don't know if this is dangerous or not, and that is where it became very murky for me.

There are many things about this short reflection that reveal the difficulties in least-worst first decision-making. Maj. Webb was faced with a dichotomous outcome with high risk on both sides. If he did not "shoot first and ask questions later," then a possible human-borne improvised explosive device (IED) could have plowed into his unit and killed multiple soldiers. If he had drawn his weapon and fired at the vehicle, then he could have killed a civilian and suffered the negative fallout associated with that course of action. Most fascinating is that even though it appears he made the right decision—he did not shoot, the man was not an insurgent, and no one got hurt—he is still ruminating about whether it was the right thing to do. Simply because of the high risk of insurgency in such situations, Maj. Webb continues to feel, at least in part, that it might have been the right course of action to shoot an innocent person. This suggests a certain consequence to the action he took; perhaps if the same situation arose again, he might be less likely to shoot because that was the correct choice the previous time, thus lowering his guard when an actual car-borne bomb barrels through the checkpoint. A further point is Maj. Webb's acknowledgment that his training did not prepare him for this situation. He elaborates:

> So we did some fire/no-fire scenarios in training. This is certainly more muddy than an insurgent holding a weapon, now, um, in a room, this is a

black box, literally, I can't see inside it, and I don't know what the intent is, I only have his actions to judge. Is he a reckless driver? Reckless drivers were not at all unusual in Afghanistan. His vehicle was unusual in Afghanistan. Or is this just something dangerous. Yeah we did do training, this looked nothing like the training.

Maj. Webb sheds light on something we all know: One cannot prepare for all eventualities, whether in business, sports, or the military. Although correct to acknowledge that this scenario had not emerged in training, Maj. Webb does not make a strong connection between previous fire/no-fire decision training and the current scenario. We suggest that perhaps there is more utility in simply having had to decide whether or not to fire in training than the officer suspects. Simply engaging the brain circuits that are involved in making such difficult decisions increases the likelihood that they will be an effective decision aid when the time comes to make such decisions for real. Although Maj. Webb correctly observed that he had not been trained for this particular event, he *did* have training in what it feels like to go through the machinations involved in deciding whether a situation requires using one's weapon.

This kind of rumination within our sample was not unusual. Consider, for example, Col. Barker (also outlined in Chapter 4), who, after receiving intelligence through Special Forces, launched an airstrike on a woodland compound. Although he acted on the view that the target was an insurgent camp, it was not until he received a call from his Brigade Commander that he realized there was an alternate perception and they could be civilians. Even now, Col. Barker remains unsure as to how (and why) he made the decision he made, ruminating still on the factors that led to that decision:

> And erm . . . so erm . . . so that's something you know, looking back on it for a number of years and doing some leadership and decision making courses, I don't, I can never understand why I made that choice. You know, I think it was out of desperation and also because you know I had taken some personality tests and I think I'm a very intuitive kind of thinker, maybe more so than I need to be, you know, at times.

But even with the benefit of hindsight, we can still see that there is a powerful tendency for the colonel to second-guess his decision-making and to wonder what the consequences would have been had he acted hastily on bad intelligence instead of acting hastily on good intelligence.

The previous two vignettes highlight decisions that resulted in desired outcomes for the coalition forces. In Maj. Webb's case, the decision *not* to shoot led to a positive outcome—no loss of civilian life and no threat to his unit. In the language of signal detection theory, a concept from radar signal processing that has become a key foundation of psychological theories of decision-making (Green & Swets, 1966), this is called a *correct rejection*. For Col. Barker, we see the opposite side of the cross. Acting quickly and decisively led to an attack on enemy assets that furthered the cause of the colonel's unit; in signal detection theory, this is referred to as a *hit*. But what of decisions that go wrong? For Col. Barker, had he failed to act, this would have been referred to as a *miss*—there would have been a legitimate enemy camp that was ignored. For Maj. Webb, acting would have resulted in a civilian death—in the dry language of signal detection, a *false alarm*. These kinds of negative outcomes, misses and false alarms, often form the basis of ruminative what-if thinking: "What if I had stopped that guy instead of letting him through?" (*miss*); "What if I that airstrike had hit a civilian camp?" (*false alarm*).

As you can imagine, these ruminations are even more significant in the case in which action *did* have negative actions. This is especially pronounced in those situations in which one does not know for sure that an alternate choice could have resulted in a less-worse outcome (which is really a defining factor of least-worst decisions). A poignant example of this perhaps is Capt. West, who, as discussed in Chapter 8, had to decide if (and how) to evacuate a hostile city. His decision resulted in his soldiers stepping on embedded IEDs and significant harm. As you can imagine, his ruminations are even more pronounced that those of Maj. Webb and Col. Barker:

> I think about this all the time, and I came out without a scratch and it's one of those things those three guys, or four guys, who were wounded almost instantaneously, you think you know, maybe I could've made a different decision? But those are alternate realities.

Ruminations such as this have been shown to be a strong predictor of a PTSD diagnosis, with some researchers delving further and showing specifically that the tendency to ask "What if?" questions, the compulsion to continue ruminating, and the occurrence of unproductive thoughts are the facets of rumination that are most strongly associated with PTSD both cross-sectionally (in individuals who have PTSD vs. those who do not have PTSD currently) and prospectively (in individuals who will be later diagnosed with PTSD vs. those who will not) (Michael et al., 2007). However, before discussing how least-worst decision-making may leave soldiers particularly susceptible to PTSD and also the therapeutic approaches to treating it, we briefly discuss the history of PTSD. After which we turn to its prevalence, the factors that contribute to and protect from it, and a discussion of the brain networks involved in PTSD.

POST-TRAUMATIC STRESS DISORDER

Post-traumatic stress disorder has been named since 1980, when the American Psychiatric Association added the diagnosis to the third edition of its *Diagnostic and Statistical Manual of Mental Disorders*. Although initially controversial, the diagnosis has gained widespread acceptance throughout the years due to the increased light shed on the role of traumatic events and the negative experiences of the survivors of such events. This diagnosis shifted the emphasis away from factors internal to the sufferer toward the precipitating traumatic events, reifying the notion that PTSD is properly understood only in the context of the trauma from which it arises. In the military context, PTSD has been observed since humans began writing about battles, with Shakespeare in *Henry IV, Part 1* famously describing the nocturnal stirrings of Hotspur as a constant rehashing of the battles in which he previously engaged (Bennet, 2011). What was previously referred to variously as *shell shock* or *combat exhaustion* (Friedman, Schnurr, & McDonagh-Coyle, 1994) was now viewed as a disorder that occurred in response to a traumatic precipitating event of many kinds, not just war. However, the prevalence of PTSD in post-combat soldiers relative to the general population appears to be much higher. The National Vietnam Veterans Readjustment Study (NVVRS; Kulka et al., 1990) estimated the lifetime prevalence of PTSD in Vietnam War veterans to be approximately 29% (averaged across men and women), whereas a nationally representative survey in the United States assessed lifetime prevalence to be approximately 6.8% (Kessler et al., 2005). Estimates of prevalence of PTSD in veterans from recent conflicts, such as Operations Enduring Freedom (in Afghanistan) and Iraqi Freedom, placed that figure at approximately 13.8%, although this figure was for current prevalence—lifetime prevalence is likely higher (Tanielian & Jaycox, 2008). Such figures serve to reinforce just how negatively charged is the experience of war and make it incumbent upon us to better understand how war foments PTSD and how we can mitigate against its occurrence.

A PTSD diagnosis is contingent upon symptoms experienced across three domains. The first, and most critical, concerns the involuntary re-experiencing of the trauma. This may take the form of nightmares, intrusive thoughts, or indeed constant rumination via what-if questions. The key point is that such thoughts are unwanted, difficult to prevent, and automatic, in the sense that they emerge in consciousness without any forewarning. The second cluster of symptoms concerns the avoidance of reminders and a numbing of responsivity. Such symptoms may manifest as an inability to love or a generally flattened *affect*, the psychological term for emotional experience. Finally, a diagnosis of PTSD can only be confirmed if the patient also suffers from increased arousal. Such physiological arousal is often manifested as difficulty sleeping or concentrating, an extreme sense of vigilance, and an exaggerated startle response, as is typically seen in fictional portrayals of PTSD. Together, these symptoms comprise a cognitive–emotional–behavioral axis of disordered reactions that vary in their extremity across both time and individuals and that can be very difficult to overcome.

We noted that combat exposure is not a guarantee that a soldier will get PTSD, so we must consider the combination of other factors that best predict who will most likely develop PTSD. Friedman et al. (1994), in their summary of what was known at the time about such factors, showed that pre-military, military, and post-military factors together accounted for approximately 40% of the variance in who gets PTSD, with genetic factors (as established in twin studies) accounting for another 30% of the variance. Such numbers show that the relationship between trauma and PTSD is complex: If we were to build a model to predict which soldiers would develop PTSD in response to war, we would need to include factors that account for (1) personality and family history, (2) the presence of abuse or neglect during childhood, (3) genetic makeup, (4) exposure to combat, (5) injury status during combat, (6) prisoner of war status during combat, and (7) social support system post the traumatic event, among others. In particular, there is evidence from a sample matched on length of service, rank, and military occupational specialty that being injured during combat makes one much more likely to develop PTSD (Koren et al., 2005). Meta-analysis (a study of many studies) can be a powerful method to assess whether factors that are thought to predict PTSD in fact do so. Two noteworthy meta-analyses highlight the roles of the following in predicting a later PTSD diagnosis: lack of social support for veterans (Brewin, Andrews, & Valentine, 2000); the perception that one's life is in danger during the traumatic event (i.e., being in extremis); and extreme emotional responses such as fear, helplessness, horror, guilt, and shame during the traumatic event (Ozer et al., 2003).

The strongest predictor of later PTSD was the experience of "peritraumatic dissociation"—often experienced as an altered sense of self, or perhaps the sensation that time is moving more quickly or slowly than usual. Such peritraumatic dissociations are similar in nature to the experience of "flashbulb memories"—vivid, hyperreal recollections of significant personal events. Some readers will be able to recall where they were and what they were doing when they heard the news of the assassination of John F. Kennedy; other readers will be able to recall a significant number of details about the events of their own experience during the September 11, 2001, attacks on the United States. Peritraumatic dissociation and flashbulb memories are similar in their unusual vividness, both when experiencing the event and when recalling it.

Work in human brain imaging using functional magnetic resonance imaging has repeatedly shown the involvement of the amygdala—a small, almond-shaped region of the brain responsible for much of our emotional experience and for signaling the presence of important, salient events in the environment in flashbulb memories. This work also implicates the hippocampus, a brain structure known to be essential for the formation of new memories since the time of patient HM in the 1950s, who had his hippocampi (on both sides of the brain) removed as part of surgery to relieve epilepsy. The unforeseen and highly unfortunate consequence of this surgery was that HM was no longer able to form new episodic memories—memories for personally experienced events. Thus, flashbulb memories appear to form as a result of the interaction of parts of the brain networks that deal with emotion and memory. The suggestion is that the strong emotional experience of the event leads to deeper encoding of all the personal details surrounding us as the event unfolds. Flashbulb memories are interesting for a number of reasons, including the fact that they are typically associated with very high confidence but poor accuracy and rapid forgetting of the relevant details. A study by a consortium of neuroscientists after the September 11, 2001, attacks revealed that flashbulb memories decay rapidly (after approximately 1 year) but then remain stable at least 10 years after the event. As previously mentioned, however, people's confidence in their flashbulb memories remained very high throughout. It is as if the emotional brain network's "tagging" of the event's importance leads us to believe in our memories more, even when they no longer accurately reflect the events we lived through. Mirroring this, much research in neuroimaging also implicates the amygdala and hippocampus in the experiencing of PTSD symptoms.

Current thinking about the brain representation of PTSD suggests that the amygdala is hyperreactive in PTSD. This can manifest itself as increased amygdala activation when PTSD sufferers hear combat sounds (Liberzon et al., 1999) or encounter trauma-related words (Protopescu et al., 2005), for example. The importance of this exaggerated amygdala response is that one of the functions of the amygdala is to help learn conditioned fear responses; in some sense, the amygdala serves as an alarm bell, signaling that the particular environmental conditions should be making us afraid. The amygdala is also, at least in part, necessary for the startle response (Angrilli et al., 1996). Because PTSD sufferers show exaggerated amygdala responses, it follows that they will also likely generalize their fear to stimuli that are not in themselves fear-inducing. Such a negative feedback loop may ultimately make the situation worse for sufferers of PTSD, as they overgeneralize their fear triggers to previously innocuous environmental events (e.g., the sound of a car door slamming). In most individuals, the amygdala is part of a neural circuit that is attuned to important environmental events, and it is subject to being inhibited ("downregulation") by other brain regions such as the medial prefrontal cortex in the frontal lobes. This means that we have a neural circuit that "learns" associations between environmental events and their importance to us. The amygdala may act as an alarm signal to warn us, but the medial prefrontal cortex—which inhibits the amygdala—receives projections from other brain regions that are responsible for making meaning out of environmental events. Such information processing allows for a dampening of the amygdala response both acutely and chronically. If you jump at seeing a shadow moving on the wall and then realize it is your own, your first response is to immediately calm down, and the next time it happens, your amygdala will be less likely to ring the alarm. This feedback loop appears to be compromised in PTSD, with researchers finding reduced cortical volume (likely fewer neurons) in the medial prefrontal regions responsible for attenuating amygdala responses (Rauch et al., 2003). Importantly, such reductions are observed when comparing PTSD patients to people who have witnessed trauma but did not develop PTSD—a very strong control indicating there is something specific about the PTSD response to trauma and not just the trauma itself that involves a reduction in medial prefrontal volume.

Finally, we also observe that the hippocampus typically presents with reduced volume in PTSD sufferers compared to trauma-exposed controls who did not develop PTSD (Gurvits et al., 1996). Not only is there reduced volume but also this decrease in volume is correlated with increased PTSD symptoms. This reduction in hippocampus volume suggests difficulties in encoding novel memories, which is one of the techniques that therapists use to help PTSD sufferers overcome the disorder. With an impairment in hippocampal function, it follows that replacing the negative, traumatic memories with new, positive memories will be a more difficult task. In fact, a 2016 study on the brains of PTSD patients compared to those of trauma-exposed healthy individuals showed that of the PTSD sufferers who responded well to a 10-week course of exposure therapy, there was a significant increase in the volume of the hippocampus (Rubin et al., 2016). These data suggest that this increase in hippocampus volume might have been associated with an improvement in the ability to create fear-extinction memories (i.e., to replace negative memories with neutral ones). This is an exciting result, suggesting a potential mechanism for the reduction of PTSD, but it is certainly in need of replication in larger samples.

The neuroscientific investigation of PTSD is ongoing, with much to learn about how the brain processes information in PTSD sufferers and also much to learn about whether the neural differences observed thus far occur as a result of the trauma, predate the trauma and thus make sufferers more susceptible, or even occur as a result of changes that soldiers go through after returning home from war, such as difficulty readjusting to everyday life. For now, we note that this is a highly topical and burgeoning research area, with likely many interesting and helpful findings emerging to aid in the treatment of PTSD.

We previously noted that meta-analyses have shown that the factors surrounding the traumatic event have quite a large effect on PTSD. Other factors outside of the traumatic event, such as prior adjustment, prior traumatic experiences, and a family history of psychopathology, also have smaller but not negligible effects on the prediction of PTSD (Ozer et al., 2003). A final note concerns gender. Several reviews demonstrate that when women and men are exposed to the same trauma, or when the kind of traumatic event is considered in the statistical model, there is much greater prevalence of ensuing PTSD diagnosis in women; in fact, women are twice as likely as men to develop PTSD (Norris, 1992; Wolfe & Kimerling, 1997). Although women are disproportionately the victims of sexual violence, their rates of PTSD are still double those of men, even when the rates of traumatic abuse are taken into account. Such statistics are critical to consider as military occupation specialties are currently being opened to women in the US Armed Forces. Open questions remain regarding the role that gender plays in PTSD development, given that those members of the military who serve on the front lines, in the infantry, are those exposed to the greatest trauma and the most likely to suffer from PTSD (McNally, 2003).

But even by bringing all of this knowledge to the table, and by including measurement of all of these factors in our statistical models, we would achieve the right answer approximately 70% of the time. So, there is still much to learn about what experiences—pre-, during, and post-war—and which individual difference factors can help us understand who is most likely to go on to develop PTSD in response to combat deployment. As mentioned previously, digging deeper into the nature of combat exposure, determining who had to make least-worst first decisions versus who did not have to make such decisions might help us better specify why some people get PTSD and some do not. Why is this question important? One reason for a premium on predicting who will get PTSD is that we could use such information to screen people prior to deployment: If we can derive from a model a relatively accurate probability of getting PTSD, then that could allow us to make early decisions about who to assign to which military occupation to reduce their likelihood of having to deploy.

MORALLY INJURIOUS EXPERIENCES

In an effort to further understand the link between what happens at war and the likelihood that a returning soldier will suffer from PTSD, researchers have increasingly begun to examine events termed as "morally injurious." Such events are a separate category from experiencing trauma (i.e., being in a firefight, shot at, or involved in a IED event) and instead involve witnessing or engaging in acts that transgress our own morals or beliefs. The result of experiencing such events is then referred to as a morally injurious experience (MIE), which reflects the guilt, shame, anger, self-handicapping behaviors, and relational problems (e.g., social alienation) that can emerge after witnessing and/or participating in events that "challenge one's basic sense of humanity" (Currier, Holland, & Malott, 2014, p. 3). MIEs betray "what's right" in a high-stakes situation (Shay, 1995). Thus, the likelihood of PTSD is closely linked to moral injuries as much as it is linked to physical injuries.

As MIEs gain interest in the research world, it is increasingly being shown that exposure to MIEs is closely associated with experiencing PTSD on return. For example, when controlling for being involved in life-threatening events during deployment, war-zone service studies have found that exposure to morally injurious stressors is associated with PTSD and other reintegration problems (e.g., substance misuse and suicidal thoughts/behaviors; Currier, Holland, Drescher, & Foy, 2015). Currier and colleagues (2015) sought to identify the prevalence of MIEs in veterans with PTSD. They interviewed 14 male veterans in a PTSD recovery program on the West Coast, asking them the following:

Interviewer: During your deployment, did you confront any ethical dilemmas? Were you confronted with any situations in the war-zone in which the correct decision or action was unclear? If so, please describe these experiences.

Currier and colleagues then coded the veterans' responses to measure the different types of MIEs to which they were exposed. All 14 veterans said that what they "saw/experienced in the war left [them] feeling betrayed or let-down by military/political leaders." More interesting, perhaps, is that in addition to seeing acts that violated their moral principles, 11 of the veterans (approximately 80% of the sample) reported that they "did things in the war that betrayed my personal values." Even more relevant, perhaps, is that 11 of the veterans also reported that they "had to make decisions in the war at times when I didn't know the right thing to do."

MIEs therefore offer a suitable avenue through which we can explain the link between least-worst decision-making and issues with reintegration and PTSD. For example, making decisions when "you didn't know the right thing to do" is a metric on the Moral Injury Questionnaire–Military Version (MIQ-M; for an outline of the MIQ-M, see Currier et al., 2015), which is now extensively used in veteran populations (Farnsworth et al., 2014). Furthermore, given that not "doing what's right" and doing things that "transgress deeply held moral beliefs and expectations" (Litz et al., 2009, p. 700) are central to MIEs, least-worst decisions are likely to factor in many MIE-inducing situations. Furthermore, and observed in Chapter 5, least-worst decisions are closely linked to (and often driven by) the values that people hold sacred (e.g., force protection). Making least-worst decisions might therefore require individuals to trade off against two (potentially) sacred values, meaning that (no matter which one they choose) they must act in a way that violates one or more of the morals and values that they deem important—a central aspect of MIEs. Our observations of soldiers ruminating on such decisions, coupled with the very strong memory traces they have for these events, thus provide tacit support for the long-lasting effect of such decisions.

However, it is important that we do not draw too strong a causal relationship between the two because although Currier and colleagues (2014, 2015) found high levels of MIEs in veteran samples with PTSD, this does not mean that MIEs (and hence making least-worst decisions) will result in someone developing PTSD. We also found that many of the individuals we interviewed who had made least-worst decisions did not have trouble integrating, and some demonstrated an amazing resilience. One of our interviewees, an Air Force pilot, explained his thinking on rumination:

> I think I would have made the same calls . . . I can't think of too many situations that I was in where I looked back and said "I wish I would have done that differently" but that's kind of my personality anyway. I don't spend a lot of time on . . . "would of, could of, should of" kind of stuff. . . . I personally think that makes a good aviator kinda like it makes a good goalie in football or makes a good pitcher in baseball. I mean especially a relief pitcher you come in a throw 100 balls and 9 times out of 10 your successful and you dwell on the 1 time out of 10 that the guy rocks one out of the fence for a game winner? You know you're gonna have a hard time so . . . you know I try to use all the mistakes that I made flying to get better but I didn't dwell on them.

RESILIENCE TO PTSD

Thus far, we have mainly discussed the nature of combat experience, the broad role of genetics, and early childhood experiences in predicting who will develop PTSD. But what of the factors that may protect against PTSD? Some clues have come from the literature on intimate partner violence. In particular, a study by Coker and colleagues (2005) showed that the following factors appeared to increase resiliency of individuals who had been subjected to intimate partner violence (i.e., those who did not later develop PTSD): higher education and income, being currently married, and reporting that the intimate partner violence had stopped. Extrapolating to the military combat domain, we can easily see that there may be a role for education and relationship status in protecting against the development of PTSD. Renshaw (2011) developed an integrated model of the factors that might help protect against the emergence of post-deployment symptoms for veterans of the recent conflicts in Afghanistan and Iraq. One major factor he highlighted in his review of protective factors was the effect of pre-deployment preparation and training specifically for combat exposure. Renshaw noted that the *perception* of threat to life might mediate the relationship between combat exposure and later PTSD. In other words, although there is a relationship between combat exposure and later PTSD—more combat exposure often leads to more PTSD—this relationship can in fact be explained by soldiers' ratings of how much their lives were in danger during the combat exposure. Thus, by examining what specifically about combat exposure predicts PTSD, we have established a role for how soldiers perceive those events in explaining their later disorder. In Renshaw's analysis, his data suggest that troops who are more prepared for deployment may be less likely to perceive combat situations as overwhelmingly threatening and therefore less likely to suffer PTSD post-deployment.

Renshaw (2011) also noted a difference between combat exposure and what he termed "post-battle" experiences. Combat exposure, as shown, refers to the events that occur during combat and for which pre-deployment preparation appears protective. Post-battle experiences, on the other hand, have some overlap with the morally injurious experiences we referenced previously. These are things such as seeing severely wounded individuals. The term "post-battle" experiences refers to the period of time directly after concluding a battle rather than, for example, readjustment periods back at base or after action reviews. In the statistical model that Renshaw constructed, post-battle experiences had a direct effect on later PTSD symptoms, which was not mediated by perceptions of threat. This makes sense given that moving from the combat phase to the post-battle phase is associated with a reduction in the immediate real threat (less fear) and that the post-battle period is likely more associated with the shock and horror of war.

Such results go deeper than simply acknowledging that combat exposure can often lead to PTSD. They also suggest that being prepared for the difficulties of battle can reduce the amount of threat felt by soldiers. We have just discussed how this can protect against later PTSD, but it is worth thinking about what the mechanism of protection might be. Why would soldiers who feel less threat during combat be protected against negative, intrusive thoughts, higher physiological arousal, and a numbing of emotions months and years after the exposure? One possibility (among many, of course) is that reduced perception of threat during combat can likely lead to better decision-making during those moments. Someone who is suffering from extremely high physiological arousal, a high physiological stress response brought about by extreme fear, can make decisions that are less optimal, more risky, and more reward-seeking. Given this, it is helpful here to take a quick detour into understanding the effects of stress once again.

Stress is a physiological response to environmental events. Stressors can be out there in the world, requiring us to process them emotionally and cognitively. In the context of war, these so-called *processive* stressors might be IED explosions, multiple radio conversations, the sound of gunfire, uncertainty about enemy locations, intel requiring decisions, among others. Stressors can also impact on our body directly, causing immediate threats to homeostasis. These so-called *systemic stressors* might include extreme thirst, fatigue, open wounds, or hunger—all events regularly experienced by soldiers much more so than by the civilian population. Processive and systemic stressors can occur at the same time in daily life, and in the context of battle, they almost always occur simultaneously. Soldiers are often asked to complete missions of up to 24 hours' duration with little or no time available to sleep, often without enough food or water to sustain them for the intense physical exertions they must undergo (the effects of which we discussed in detail in Chapter 8). Thus, soldiers must deal with physical demands well beyond those experienced in civilian life, with cognitive demands that often exceed those we normally encounter; in addition, they must deal with the kinds of horrific negative events that characterize close combat. The point is that stress, as it is medically defined, is a constant in the life of a deployed soldier.

So how does stress affect decision-making? A 2016 meta-analysis by Starcke and Brand (2016) showed that stress conditions (in the lab) led to decisions that were more reward-seeking and more risk-taking. Furthermore, such effects were strongest for decisions for which risk-taking and reward-seeking were bad ideas, and in studies in which the authors characterized the kind of stressor they were investigating, effects were strongest for processive rather than systemic stressors. It is worth reiterating that these conditions prevailed in a lab setting. We do not yet know how these effects might play out in a military setting. But we can make a good guess that if lab stressors (e.g., being asked to make a speech in front of an unfriendly crowd or sticking one's hand in ice water) are enough to negatively impact decision-making, then it is likely that military operational stressors will further negatively impact decision-making. Unpacking these lab results, it can be seen that in high-risk circumstances, in which risk-seeking behavior is a bad idea, having to deal with processive stressors in our environments can lead us into increased risk-taking and reward-seeking—behaviors that are disadvantageous in the kinds of tasks tested in these lab protocols. Because soldiers are often placed under conditions of high processive stress and high risk, it follows that their decision-making may be negatively affected. Col. Barker appears to represent a classic example of a commander having to deal with multiple environmental stressors engaging in behavior that is high-risk and high-reward, without thinking clearly through all possible courses of action. It also follows that these very decisions are the sorts that will be later ruminated over; once the fog of battle has cleared and cooler cognitions emerge, the processive stressors that push us in one direction no longer hold sway. It is then that we can start to understand *why* the decisions we made are potentially suboptimal, and it is also at this point that we might start to engage in the kind of ruminative what-if thinking that characterizes the mental lives of PTSD sufferers.

Even without considering the numerous effects of physical stress, we have seen that soldiers are consistently placed in situations in which the factors they must deal with combine to create the likeliest possible conditions for them to develop PTSD. High stress, in contexts with no good outcomes, with no prior exposure, and the constant threat of injury or death, makes for conditions in which resiliency and top-notch decision-making are at an absolute premium but in which cross-sectional retrospective field studies and lab experiments have demonstrated that people simply cannot perform at their best.

What is surprising, under these circumstances, is that more soldiers do not develop PTSD upon coming home. It is clear that more studies are needed to understand why, fortunately, so many are spared. Research into the genetic factors that increase resilience, the childhood factors that make individuals more resilient, the protective factors of education and training, and how to better select soldiers for the kinds of stressors they will face is badly needed. For now, we end by noting that progress is being made; we are starting to be able to predict with some accuracy the conditions that are most likely to foment PTSD in those who come home. Clearly, a greater emphasis is needed on social support post-deployment. Soldiers who feel welcomed, who are helped to integrate into the community, who have families to return to, and who can find meaningful employment are most likely to be safe from harm. Although many private foundations exist to help veterans, the need is great and the supply is short. Perhaps more focused and sustained efforts to create positive, caring infrastructure for returning veterans will be a goal of future public programs. The benefits to our whole society of engaging with our veterans this way cannot be overstated.

CONCLUSION

This book has focused on the hard decisions that soldiers have to make on deployments. We have examined what makes these decisions hard in terms of the decisions themselves and the factors within the environment (both exogenous and endogenous) that make these decisions challenging. But we would be remiss if we did not look beyond the situations themselves and the tactical and strategic consequences of their decisions and consider the long-lasting impacts that these decisions have on the soldiers who have to make them. In Capt. West's words,

> Any country I would hope has an expectation out of certain people to go and do things that they just don't want to do, and they want us to do hard things, and sacrifice and sometimes that means that I will do things that may mean that maybe I'm not as safe as the people back here, and I'm ok with it.

War is hard, and the decisions that soldiers make cause them conflict. Although this psychological concept of decision conflict is widely studied, often ignored is the long-lasting conflict that such decisions can cause and the implications this has for reintegration post-war. In this chapter, we aimed to sketch the trajectory from least-worst decisions to wider issues that the returning soldier must face, and we hope that greater awareness of this relationship can lead to greater help for our veterans.

10

HOW DO SOLDIERS DO WHAT THEY DO AND WHAT CAN WE LEARN FROM IT?

Soldiers enter the Army with their own values, developed in childhood and nurtured through experience. We are all shaped by what we have seen, what we have learned, and whom we have met. However, once Soldiers put on the uniform and take the oath, they have opted to accept a Warrior Ethos and have promised to live by Army Values. Army Values form the very identity of the Army. They are nonnegotiable and apply to everyone at all times, in all situations.
—United States Army (2014, p. 1)

It's not hard to make decisions when you know what your values are.
—Roy Disney

Throughout the course of this book, we have introduced various models of decision-making, highlighted our research on how soldiers actually make decisions in the field, and covered the negative consequences that often arise from having to select the least-worst first. In this chapter, we seek to compress these insights into an overarching understanding of the core psychology of least-worst decision-making. In addition, we look to the future. Specifically, we focus on one of the findings that struck us most as authors—the fact that soldiers, at least in one respect, often seem to outperform their non-military counterparts with regard to navigating fraught psychological terrain with ease and conviction, proving decisive where others might falter, and arriving at courses of action that maintain operational tempo. This chapter highlights what we can distill from this knowledge to (1) further our understanding of the soldier's special environment, (2) improve how soldiers and others make decisions, and (3) take lessons and apply them to bettering the lives of the 99% who ultimately do not serve.

HOW MILITARY DECISIONS ARE MADE

When reflecting on the contribution of this book to our understanding of how military decisions are made, it is perhaps prudent to re-visit Gary Klein's (1989) original article on the subject, in which he argued that the military needed to move away from analytical decision-making. In his words,

> It is time to admit that the theories and ideals of decision-making we have held over the past 25 years are inadequate and misleading, having produced unused decision aids, ineffective decision training programs and inappropriate doctrine.... The strategies sound good, but in practice they are often disappointing. They do not work under time pressure because they take too long. Even when there is enough time, they require much work and lack flexibility for handling rapidly changing field conditions.... The point for this article is that there are different ways to make decisions, analytical ways and recognitional ways, and that [we] must understand the strengths and limits of both in order to improve military decision-making. (p. 56)

Klein (1989) observed platoon leaders and battle commanders (at a series of training events) and found that "85% of the decisions were made in less than 1 minute" (p. 58). From his estimations,

> Experienced decision-makers handle approximately 50 to 80% of decisions using recognitional strategies without any effort to contrast two or more options. If we include all decision points, routine plus non-routine, the portion of RPDs goes much higher, more than 90%. (p. 59)

It is clear then that most decisions being made were not multiattribute and did not involve the comparison of multiple courses of action (as is outlined in military doctrine). In fact, as Klein reports, when he told soldiers he was studying decision-making, one replied that he had "never made any decisions." As Klein notes, "what he meant was that he never constructed two or more options and then struggled to choose the best one" (p. 58). The reader may see the parallels here with some of our own conversations with soldiers in which they felt like they had never made a decision (see Chapter 5). Now, before we move forward to the present day, it is important that we outline some of the finer points of Klein's argument, especially as they pertain to the importance of multiattribute decision-making. Despite the over-representation of recognition-primed decision-making (RPD) decisions in the military (as high as 90%), Klein still maintained the importance of considering both strategies. As he stated, "The point ... is that there are different ways to make decisions, analytical ways and recognitional ways, and that we must understand the strengths and limitations of both to improve military decision-making" (p. 57). Hence, in Klein's view, although there was an absolute need for a refocusing of training and doctrine to emphasize recognitional methods, this was not to be done in lieu of multiattribute methods. There simply needed to be a balance between the two because both have strengths and weaknesses.

From here, then, there has been a significant push toward developing this recognitional ability (sometimes referred to as "intuition"; see Peterson & Seligman, 2004). As Mike Matthews outlines in his "vision" for training intuition in soldiers in 2030,

> To use Klein's naturalistic decision-making terms, our fictional recruit is building a sophisticated and detailed inventory of experiences—as real as those he will later build through real world missions—that will allow him to (bloodlessly, as General Scales would say) response quickly, accurately, and effectively when his boots hit the ground in some faraway land. (p. 67)

The importance of Klein's work in developing such intuition is clear. Over time, however, the nuances of this paper have become lost, and perhaps the scales have swung too far away from multiattribute decision-making—well, not multiattribute decision-making per se (for which we outline the issues in Chapter 2), but at least the idea that people make decisions *at all*. We propose that denigrating the process of choice entirely is theoretically damaging for our understanding of military decision-making, as well as problematic if we are not training soldiers to make effective choices when the time comes, even if it is only 5% of the time. This is especially true if (as the interviews support) those 5% of choices are what Taleb (2008) calls "black swans" in that they are rare, unpredictable, high-risk, and very high consequence. If so, then it is when decision-makers most need to make decisions that their intuition is often not enough.

The first contribution of this book, then, is to re-emphasize the importance of choice within a military context. Now, at first this might sound like a calling card to bring back multiattribute decision-making; it is not. Although we maintain that members of the military do make choices, these choices are not those that can be explained with multiattribute theory (for the many reasons outlined in Chapters 2 and 3). Instead, we posit that members of the military make choices that are least-worst choices, and these choices do require a process of option evaluation and eventually "choosing" a "better" choice. It was here that, in Chapter 5, we offered the important role of value systems as a way to understand how individuals differentiate between equally unappealing options.

This book has also highlighted the importance of considering the ecological niche within which such decisions are made. Although multiattribute perspectives view decisions as the outcome of a comparison of the merits of the available options, the narratives we outlined here provide a far richer and far more complex picture (as is the goal of naturalistic decision-making in general). Here, our data highlight the importance of exogenous and endogenous uncertainty, trust, anticipatory regret, role confusion, inter-agency cooperation, and interpersonal and interorganizational conflict. Hence, as stated in Chapter 3, this book strongly supports the relevance of the SAFE-T model as a knowledge-generating platform for military decision-making as it pertains to the detrimental effect of these many external and internal pressures on the process of decision-making. Although here we have used the SAFE-T model as a framework to work our way through the complex process of making decisions, we argue that as a model, it, above all others, accommodates the diverse range of influences that impact military decision-makers.

Psychological research has extensively tested the presumptions of RPD (e.g., experience and situational awareness) in military decision-making (Baber, Fulthorpe, & Houghton, 2010). Others have focused on the role of uncertainty and time pressure on military decision-making (Ahituv, Igbaria, & Sella, 1998), but many of the factors within the SAFE-T model remain unexplored. For example, no research (besides the qualitative data presented here) has examined the role of inter- or within-team trust, accountability, or anticipatory regret on military decision-making. Our data support that these are important factors and that further work needs to experimentally manipulate such factors (either lab-based or naturalistically within wider scenario-based learning) to measure their impact. Such research will have important implications not only for understanding decision-making but also for informing organizational policies such as the degree of accountability we place on soldiers during war.

LEAST-WORST DECISION-MAKING AND DECISION INERTIA

A principal premise of this thesis was to explore the psychological manifestation of decision inertia as it pertains to the military decision-making process. However, our early interviews showed some resistance to inertia within the soldiers we interviewed (or at least their recollection gave no indication of inertia). As with any scientific enquiry, our goal then became to identify *why*? What is it about those who enter the Armed Forces that enables them to commit to choices when it counts? But beyond just identifying their resistance, this book also sought to identify a theoretical explanation for a soldier's resistance to inertia. Based on our analysis of the interviews we conducted, we developed an emerging theory in Chapter 5 about the importance of values. From a process of grounded theory and theoretical deepening, we began to understand the role that values play in decision-making and, specifically, how the nature of value clashes predicted whether inertia did or did not emerge. In our interviews, when individuals were forced to choose between two values that they held equal (e.g., protecting civilian life and protecting fellow soldiers), they struggled. On the other hand, if the choice involved only one value (which we commonly noted to be the protection of fellow soldiers), decisions were often made with ease because of an unwavering lack of willingness to sacrifice against that value. Time and time again, our cases reaffirm this point in that soldiers made incredibly hard choices with relative ease because one option (and not the other) risked the life of the men and women under their command.

THE "SUBTLE ART OF NOT GIVING A F*CK"

As the final chapters of this book were being written, there was a book on the *New York Times* bestseller list that overlapped with the underlying ethos of this book (despite its less-than-scientific phraseology). In late 2016, Mark Manson released a book titled *The Subtle Art of Not Giving a F*ck*,[1] within which he provides a modern commentary on issues surrounding happiness and self-worth. In the opening section of this book is a sentiment that, to us, sums up the reflections of the hundreds of hours of interviews conducted with soldiers and, crucially, why they are better at accepting least-worst options: "The desire for more positive experience is itself a negative experience. And, paradoxically, the acceptance of one's negative experience is itself a positive experience" (p. 7). To rephrase Manson, the quest for a positive outcome creates a negative outcome, whereas the acceptance of a negative outcome creates a positive outcome. This is what the philosopher Alan Watts referred to as a "backwards law," referring to the fact that the quest for one thing can create the opposite outcome (in effect, it is the opposite of a self-fulfilling prophesy in that the belief that something is the case creates that eventuality). Putting this in perspective, it provides a frame of reference for our emerging understanding of military decision-making. What the soldiers in this study seemed able to do was to accept a negative outcome and act. On the other hand, and from a detailed analysis of Power's (2016) work and Alison's wider work in this area, it is the efforts of others to make a positive out of a potentially negative situation that creates a negative outcome (often decision inertia).

Now, let us dig deeper because it is important that we emphasize that what differentiates soldiers is not the fact that they "do not give a f*ck" (this could not be further from the truth). Manson (2016) outlines several subtleties to "not giving a f*ck" because, as he notes, "there is a name for a person who finds no emotion in anything: a

---

[1] Although we apologize for the vulgarity of the language, this is, unfortunately, the language of choice of the author. It does not reflect the views of these authors, nor does it imply, in any way, that soldiers "do not give a f*ck." As we have demonstrated throughout this book, this is not the case; and as readers shall see, this is not the argument of Manson.

psychopath ... [and] why would you want to emulate a psychopath?" (p. 13). First, "not giving a f*ck" does not mean being indifferent (i.e., caring about nothing). Second, and this is critical, "not giving a f*ck" about adversity means caring about something more important than adversity. Let us put these two subtleties together with our data. From a qualitative standpoint, soldiers demonstrated a willingness and ability to tolerate the adversity of a bad outcome to make a critical least-worst decision because they cared about something *more* than this adversity. That is, they held values that trumped any adversity they would experience because of their decision.

Although here we have applied our data to a "popular psychology" model of values and decision-making, the underlying ethos is the same: Soldiers seem to value the welfare of others over negative consequences to themselves. It is clear that to begin to apply this finding, significant future research is required to better understand this relationship, but it offers significant promise to support the role of values as a central factor in individual differences.

"THERE IS NOTHING LIKE A SOLDIER"

This quote was taken from a poem published on the website Project Afghan,[2] and it is a sentiment with which we have come to agree. To us, one fact rings true above all others: Soldiers are very good at being decisive when it really counts. Like all good scientific inquiry, answers lead to questions: Why is this the case? Why do soldiers perform better? Where does their improved decision-making come from? As with any cognitive ability, we expect, the answer is a mix of genetic endowment, life experience, personality, training, and factors in the environments in which the skill is deployed (including policy, organizational climate and culture, and one's own identity as a soldier). Beyond these clear truisms, we can attempt to zero in on which of these factors are different in soldiers. We should be clear that the research is very far from being complete. For example, almost nothing is currently known about the roles played by variations among individual genes in how people behave. But what little we currently know about all these factors can give us clues as to where we should look next.

HOW DO SOLDIERS DIFFER FROM CIVILIANS ON PERSONALITY?

Psychologists generally speak of personality varying along dimensions that are roughly independent, known as the "big five" factors: extraversion, neuroticism,

---

[2] Project Afghan is a blog written by an Army soldier who sought to capture 6 months of being at war. It can be found here at https://projectafghan.wordpress.com.

openness to experience, conscientiousness, and agreeableness. These five factors of personality are thought to account for a large amount of variation in the general population, and we can see how specific combinations of factors might cluster in specific job functions. For example, a prototypical librarian would be highly conscientious, meaning he is organized, takes pride in his work, and is consistently diligent. Our stereotypical librarian might also be introverted, perhaps preferring the company of books rather than people. A pilot might be very low on neuroticism (i.e., high on emotional stability), and therefore calm in the face of adversity, and high on openness to experience—a trait indicating he or she might like to travel, try new hobbies, and so on. Finally, a successful sales representative might be high on agreeableness, cited as a friendly and caring person who enjoys connecting with her customers. Conversely, someone low on agreeableness would be viewed as perhaps competitive, challenging, or analytical. With regard to our interviewees, for example, several stated that they (and, in fact, "most military personnel") were a type A personality. Type A personalities are competitive, impatient, and have excessive drive. They are also more likely to engage in risky behaviors (Nabi et al., 2005). Our interviewees often attributed their confidence in their decision-making to their type A personality. Capt. Scott, for example, who, in Chapter 5, disobeyed a direct order so that he could safely return to base with his soldiers, stated,

> I'm a pretty confident guy and I think military officers in general are pretty type A and I think that you know you don't get to where you're at unless you're successful and have some amount of confidence. So I was 90% sure that . . . going up there with the QRF [quick reaction force] was the right choice and I was 110% sure that getting the hell out of there was the right choice . . . so I was pretty confident.

Col. Barker, who, as outlined in Chapter 4, was perhaps overconfident in his situation awareness, also stated that he was "type A":

> I'll say probably and to some degree arrogance factored into it . . . I think a lot of type A personalities tend to be successful in the military but also happen to be arrogant. It can be good, it can be bad, but that certainly does factor in.

Against this backdrop of—clearly stereotypical—"types" for specific job functions is the perennial chicken-and-egg question: Do people select into particular professions because they have particular personalities, or do training, experience, and expectations drive people to express themselves in ways that match those jobs? In work published in 2012, researchers sought to answer this question for a military sample. Working with German enlisted soldiers, Jackson, Thoemmes, Jonkmann, Lüdtke, and Trautwein (2012) investigated (1) the personality profile that predicts who will sign up for military service and (2) changes in personality as a result of military service. Answering this question requires a longitudinal design, whereby researchers follow individuals before, during, and after military service, and self-selection, whereby people are free to choose whether to engage in military service. Fortunately for the researchers, German law dictates that able, post-high school men must perform national service, but they are free to choose between volunteer work in hospitals, for example, and military service. This creates the ideal conditions to determine whether specific personalities predict military service opt-in and whether such service changes these individuals.

Jackson et al. (2012) found that individuals who were less neurotic (more emotionally stable), less open to experience, and less agreeable were more likely to sign up for military service. The effect sizes (a statistical term indicating, unsurprisingly, *how large* of an effect one variable has on another) were quite small, meaning that these variables did not have a major influence in predicting who served versus who did not. However, extrapolated across the population at large, these effects would not be negligible and would thus, on average, be useful in prospectively sorting people into those who will later serve and those who will not. Such results show that soldiers are slightly different from the civilian population even before they enlist.

Next, Jackson et al. (2012) examined how military training further affects personality. They found that training led to measureable reductions in agreeableness against a backdrop of general increases during that period of development (approximately ages 19–21 years): Soldiers did not increase in agreeableness as much as the young adults in the control groups. Such reductions imply that soldiers are slightly less likely to value getting along with others, less trusting, and have a less optimistic view of human nature. It is important to remember that these differences in agreeableness are present before soldiers enlist and increase as a result of military training. In fact, military leaders *higher* on agreeableness were rated as less likely to possess the transformational leadership skills necessary to identify needed change and then enact that change (Lim & Ployhart, 2004). Further evidence supports this individual finding. Costa and McCrae (1992) found among 1,300 US Air Force training pilots that their level of agreeableness was lower than that of the general population. Similarly, Klee and Renner (2016) found in a sample of 236 soldiers that they were higher in emotional stability and lower in openness and agreeableness compared to civilians.

In summary, openness (lower), emotional stability (higher), and agreeableness (lower) differentiate soldiers from civilians, and those perceived as successful leaders are also lower in agreeableness. Perhaps this is one trait that might predict who can successfully navigate least-worst first decision-making. An interesting parallel comes from research on moral psychology, in which theorists divide moral behavior into deontological (one must do no harm, regardless of ultimate consequence) or utilitarian (moral acts are justifiable based on the eventual outcome) (Greene, 2014). Deontological reasoning, the argument goes, leads to people being unwilling to endorse actions that cause harm, even when that harm leads to a greater good. Deontological judgments are characterized by emotion winning out: "I can't bring myself to cause the death of another; it's too up close and personal." Utilitarian judgments, on the other hand, are more cold, calculating, and cognitive: "If I kill one person to ultimately save five, then I've done the right thing." We all have had experiences in which cognitive and emotional considerations compete in our decisions: Do we go with our head or our heart? Sometimes we know the right thing to do but know that it will cause someone else emotional pain if we do it.

This kind of moral calculus is at the heart of least-worst decision-making. Soldiers must decide between courses of action that will cause harm no matter what is decided. Moral decision-making theorists argue that deontological and utilitarian processes are at work in tandem in all moral decisions; different brain networks contribute an emotional response and a cognitive calculation, and the winning process determines what the person will do. If soldiers are lower in agreeableness compared to civilians—and this implies that they are more competitive and analytical—then it is likely that they will be better equipped to face least-worst decisions because these decisions are akin to the utilitarian problem posed by moral ethics psychologists: Following this course of action will result in harm, but it is going to lead to the *least* harm. Coupled with this, their increased emotional stability likely implies a reduced signal from the brain networks governing emotion. Neuroscientists studying utilitarian behavior have shown that the brain regions recruited by these judgments are those involved in cognitive control (Greene, 2014), a mental function designed to help us behave appropriately in the face of environmental or personal pressures to do otherwise. Thus, the view from moral psychology might be that utilitarian judgments, those that result in the greater good regardless of causing smaller harms, are just the kinds of judgments that are required of soldiers to navigate least-worst outcome environments. In turn, soldiers' lower agreeableness, both initially when enlisting and later after military training, and higher emotional stability might together contribute to their ability to make these kinds of decisions in the face of terrible outcomes.

HOW DOES ARMED FORCES CULTURE DIFFER FROM CIVILIAN CULTURE?

What, if anything, is special about soldiers relative to people in other professions? What characteristics of soldiers' environments differentiate them? Is it character, training, or culture? We have already established that personality differentiates soldiers from civilians, both pre-service and after training. Another area that stands out is soldier culture. Soldiers must live in very close proximity to one another and uphold the Army values of loyalty, duty, respect, selfless service, honor, integrity, and personal courage. The typical soldier mission, both in training and in theater, might take 12–30 hours, with little opportunity for recovery or rest and almost no time to attend to one's own needs rather than those of the group. Basic combat training, the rite of initiation for all soldiers, is designed to foster a group mentality at the expense of the individual. It is not uncommon, for example, for an entire platoon to be punished because one soldier did not sweep under his or her bed appropriately (Faris, 1975). During combat, it is argued that soldiers are motivated to fight by "regard for their comrades, respect for their leaders, concern for their own reputation, and an urge to contribute to the success of the group" (Grossman, 1996, p. 90). All these factors together conspire to make soldier life different from that of the emergency services that respond to critical incidents; soldiers live and work in more tightly knit units, in larger groups, with greater everyday collective consequences for their actions compared to those in the emergency services. Although least-worst decisions of all stripes are characterized by exclusively negative paths, the difference is that soldiers are immersed in a culture that specifies decisive selfless action on behalf of their comrades. In contrast, emergency service personnel generally uphold the key principle of "save life," and although there is doubtless concern for colleagues, the value system is more complex. Indeed, in research on "offensive" (go in and prioritize casualties) and "defensive" (stand back and wait until the zone is made safe) goals in responding to a marauding firearms event, paramedics, police, and fire service all believed that their counterpart agencies shared the same value systems as them, even though in reality they did not. For example, paramedics were more keen to commit than police officers and firefighters in the early stages. Firefighters were more keen to commit in later stages, and police officers oscillated between the two (varying offensive and defensive action plans). Thus, not only did emergency service personnel hold different perspectives from those of members of the other services but also they assumed that members of the other agencies shared their views (when they did not—they in fact changed offensive and defensive positions throughout).

A further difference is that military decisions (if only at the small-unit level) rarely face the level of public scrutiny to which critical incident emergency response decisions are subjected. The actions of a platoon sergeant in navigating a cultural impasse in the mountains of Afghanistan are less likely to be publicly dissected in comparison to the decisions made by, for example, the Boston police chief when attempting to apprehend the Boston Marathon bombers in 2013 or police and counter-terrorism forces seeking the perpetrators of terrorist attacks in Brussels or Mumbai, where the eye of the media is constantly alert and headlines are being written as the events unfold. Police and others are well aware that these are career-defining (or career-ending) moments. Although decisions made by a general or a colonel on behalf of a large unit such as a division or brigade might be more publicly visible, such decisions are less likely to be characterized by exclusively negative outcomes and also occur less frequently than those made by the team, squad, and platoon leaders who comprise the engine of the fighting force.

FUTURE RESEARCH QUESTIONS: STUDYING MILITARY

The research we performed for this book, along with the existing military research literature, leads to numerous conclusions about how soldiers succeed in least-worst decision-making, the culture that enables this success, and the ensuing effects of having to make such decisions in theater. The SAFE-T model we highlighted goes a long way toward understanding how soldiers make decisions in real life, the inertia traps to which they might be susceptible (and what "inoculates" them against these traps), as well as prescribing how decisions *should* be made. But what can we do with this research?

One of the authors (Dr. Moran) is a cognitive scientist at the US Army Natick Soldier Research, Development and Engineering Center (NSRDEC), where he is a member of a team that conducts basic and applied research on many aspects of cognition that have relevance for soldiers. Cognitive science research at NSRDEC follows the basic construct of monitor, predict, optimize. In essence, researchers aim for a model whereby they identify a soldier-relevant situation, learn how to monitor performance and cognition in that situation, extract ground truths to predict future behavior, and finally hint at interventions that enable scientists and technologists to optimize how soldiers perform. Such a research endeavor is not unique to the Army. The Air Force Research Laboratory's 711th Human Performance Wing, for example, organizes its research goals around the concepts of sense, assess, and augment—ideas that are similar to those espoused in the Army. These approaches, coupled with the recent adoption of the Soldier and Small Unit Performance Optimization (S2PO) strategy by the Army's Research and Development Command, are designed to provide both materiel and non-materiel solutions to problems encountered by the modern warfighter. Briefly, materiel solutions are those that involve some piece of equipment, be it an ultramodern head-mounted display with real-time information or a simple stand to support the weight of gear as paratroopers prepare to board an aircraft. Materiel solutions have long been the engine of military research and design, often spilling out into the commercial sector as viable consumer products (e.g., the internet or global positioning systems). Non-materiel solutions, such as new selection and training ideas derived from applying psychological research, have the advantages of generally being cheaper and easier to field. Clearly, one of the problems that soldiers deal with is regular decision-making under uncertainty, and military researchers apply both materiel and non-materiel solutions in an attempt to solve such problems. A recurring theme of the interviews conducted for this book was that soldiers faced situations on deployment that they did not train for nor which standard operating procedures (SOPs) fully fit. Consider Lt. Col. Davids, who had to navigate the recovery of a large asset that had been ambushed by insurgents (the "large military distraction," as he called it). Although there were aspects of the *process* of recovering the asset that were covered by SOPs and training, he had far less help from training and SOPs for making the decision itself (Do I risk my men to go outside the wire and attempt a recovery?) and the actual *doing* of the recovery:

> The SOPs that we used were probably the risk mitigation measures, setting up the cordons, that would have been a SOP for anything, but for an infantry unit to go in and do some operations. Having the air coverage was probably a SOP, the movement out there was a SOP, but as far as the recovery and stuff that was kind of a new, like I said, because of the size and location, that where we really relied on the experience of the Non-Commissioned officers and the subject matter experts to figure out how they were going to solve the enigma of getting that out of there. Getting in there, clearing the area and securing the area and getting back out was all, we used SOPs for them, that is stuff that they have all been trained on before, they knew. The communications were pretty standard. I would say a lot of it was SOP, uh, the uh, decision to get in there, the decision was different, and then the actions on site kind of had to be improvised and adapted to fit.

Given the RPD emphasis on decision quality being improved by having a reservoir of similar experiences, it is obvious that disconnects between a soldier's individual training and wartime experiences will arise. Through writing this book, we have gained an appreciation that what may seem vanishingly rare through the eyes of the individual soldier may in fact be a commonplace experience when viewed through the lens of the military as a whole. With this perspective, it is possible to imagine a research program that successfully determines how to train soldiers in pre-deployment conditions to be able to use RPD when faced with uncommon yet critical least-worst first situations during their deployment. This research would fit squarely within the predict and optimize phases of research outlined previously.

The first step in any such program would involve qualitative research designed to develop a battery of the most commonly experienced unlikely events during deployment. This could be achieved via either one-on-one or focus-group-style interviews with soldiers, designed to elicit their specific recollections of events that they believed they were unprepared to deal with because of a lack of exposure during their training. Once a database of these situations is developed, it would then be possible to comb through these situations to identify commonalities. We are currently undertaking work to create this "lexis of critical incidents" (or LOCI, as we have called it in the emergency service arena) and use it to develop unlikely but challenging scenarios. The objective is to regularly expose personnel to challenges that seem real but do not actually have consequences so that we might encourage self-reflective learning, provide feedback on performance, and, ultimately, help accelerate expertise. And there was evidence that some of our soldiers exposed each other to challenging least-worst scenarios, but this was done in a more informal manner:

> I mean, for our commanders. We put hard, unsolvable problems in front of them that kind of caused everyone to exercise this, and then we usually throw in ethical dilemmas on top of that, and cause us to go through that.

In our soldier version, such events might include uncertainty about the affiliation of an unknown person, uncertainty about human intelligence, uncertainty about unmanned aerial vehicle feeds, and the like. It is now possible to use modern machine learning solutions to crunch a database of individual experiences into the set of factors that most commonly arises across the military experience. Using this approach, we could zero in on the most commonly experienced "unlikely" events (predict), which would allow for targeting the development of training simulations toward situations that soldiers will probably face when they deploy (optimize). As technology matures, simulated environments that make use of three-dimensional virtual reality, high-fidelity audio, and realistic weapons integration are being developed at various military research centers throughout the United States, including NSRDEC, the Mission Command Center of Excellence, and the new Synthetic Training Environment being led by the Army Research Lab's Advanced Training and Simulation Division. At NRSDEC, a virtual-reality MiniDome environment has been constructed and is used to measure how technologies designed to improve situational awareness impact on soldier performance and cognitive workload. These solutions, fashioned to improve soldier training and the usability of novel technologies, hold great promise when combined with the hard-won knowledge of a century of psychological investigation about how decisions are made and made successfully—an effective union between materiel and nonmateriel approaches.

FUTURE RESEARCH QUESTIONS: EMERGENCY SERVICES

Our work with soldiers has also helped us understand the challenging world of critical incident management for the men and women who are faced with critical incidents but who work as paramedics, fire officers, or police officers. We know that emergency service personnel can be more inclined to decision inertia, and there may be some methods by which we can help them identify those vulnerabilities (and thus mitigate their effect). Indeed, we have had some success with this so far in making many command-level decision-makers simply more aware of the phenomena as well as encouraging goal-directed thinking to get them over inertia traps. Moreover, our research on emergency service personnel allows us to transfer elements (methods, lessons learned, and expertise) to military environments. Our work with emergency services has taught us a great deal about how to utilize scenario-based learning to accelerate expertise, to examine moral reasoning, and to encourage self-reflective learning and creativity in decision-making. In addition, debriefing tools developed with emergency services personnel have taught us a great deal about the complex operating environment of policing, ambulance, and fire and their effective coordination. By seeking ground truth, our colleagues in the emergency services have prevented us from going down the reductionist path of good or bad decisions and have led us toward the challenges of consequence choosing, improvisation, and ethical (and least-worst) decision-making. We still have a great deal to learn, but through careful observation and conscientious analysis, it is our firm belief that psychologists can continue to assist the men and women who take on the toughest of situations and what can seem like impossible decisions.

FINAL THOUGHTS

This book started with the observation (based on our interviews) that neither doctrinal methods nor RPD sufficiently explained how decisions were made in the field. With a dearth of research outside of these perspectives to draw on, we looked to the psychological literature on how people make decisions in critical incidents and the parallels that apply between these two contexts. From here, we sought to expand the theoretical framework. We noted that it was important to factor in a series of exogenous and endogenous factors that exist within the military decision-making environment and affect decision-making. We specifically focused on decision inertia as a pitfall of making decisions when these factors are present and when they result in having to choose between difficult options. By listening carefully to soldiers who had made critical least-worst decisions and conducting critical decision method interviews, we were able to confirm that, as predicted, within the critical incident literature, many of these factors are present within military decision-making and they do affect decision-making. However, inertia was far less prevalent than predicted. Given this, we sought to understand, from a theoretical perspective, the way in which soldiers made least-worst decisions—at least compared to what we had seen in emergency service responders and law enforcement. By doing so, perhaps we could isolate what prevented soldiers from "falling" into the inertia trap.

Our theoretical exploration brought us to the importance of considering values and the role that values play as a moderator in the process of comparing options; namely when an option involves a sacred value, it is immediately prioritized above the other (unless the second option also involves a sacred value of equal importance). From here, we framed the process of least-worst decision-making as a process of value trade-offs.

In reflecting on the interviews we conducted and drawing on the data we collected, the value system of soldiers was evidently a central component of least-worst decision-making. They demonstrated (qualitatively) an ability to prioritize values within least-worst choices. Note that we are not saying soldiers are better decision-makers compared to emergency service personnel or law enforcement; rather, they are less inclined to delay when faced with competing values and when all outcomes appear to be at least somewhat negative. These findings are preliminary, and of course our understanding is limited by the lack of experimental work done in this area (although this is something we are currently addressing in ongoing research funded by the Army Research Institute). Nonetheless, we are optimistic that this book presents a strong case for future research to integrate what is known elsewhere about value systems and decision-making (see Hanselmann & Tanner, 2008). Such a program of research would help us fully understand how soldiers make critical least-worst decisions in conditions of extreme psychological and physiological pressure.

Finally, we hope that this book has given you a sense of the type of unique challenges and decisions that soldiers face. We also hope that the structure of this book, presenting detailed immersive narratives, has offered you the opportunity to put yourself in the soldiers' shoes and reflect on how well (or poorly) you would handle the situations they are asked to handle and the choices they are asked to make. A 2014 *New York Times* article by Dexter Filkins notes that countries such as Afghanistan and Iraq are not merely a "theater of war but as a laboratory for the human condition in extremis." Whether academic, Armed Forces, or a member of the general public, we hope that this book has given you a better understanding of the human condition of decision-making in extremis.

REFERENCES

Adkins, C. L., Ravlin, E. C., & Meglino, B. M. (1996). Value congruence between co-workers and its relationship to work outcomes. *Group & Organization Management, 21*(4), 439–460.

Ahituv, N., Igbaria, M., & Sella, A. (1998). The effects of time pressure and completeness of information on decision making. *Journal of Management Information Systems, 15*(2), 153–172.

Alison, L., & Crego, J. (2008). *Policing critical incidents: Leadership and critical incident management*. Devon, UK: Willan.

Alison, L., Power, N., van den Heuvel, C., & Waring, S. (2015). A taxonomy of endogenous and exogenous uncertainty in high-risk, high-impact contexts. *Journal of Applied Psychology, 100*, 1309–1318.

Alison, L., Power, N., van den Heuvel, C., Humann, M., Palasinksi, M., & Crego, J. (2015). Decision inertia: Deciding between least-worst outcomes in emergency responses to disasters. *Journal of Occupational and Organisational Psychology, 88*(2), 295–321.

Alison, L., van den Heuvel, C., Waring, S., Crego, J., Power, N., Long, A., & O'Hara, T. (2013). Immersive simulated learning environments for researching critical incidents: A knowledge synthesis of the literature and experience of studying high risk strategic decision making. *Journal of Cognitive Engineering and Decision Making, 7*(3), 255–272.

Allen, C. D., Coates, B. E., & Woods, G. J., III. (2012). *Strategic decision making paradigms: A primer for senior leaders*. Carlisle Barracks, PA: Army War College.

Allen, C. D., & Gerras, S. J. (2009). Developing creative and critical thinkers. *Military Review, 89*(6), 77–83.

Alper, M. (2014). War on Instagram: Framing conflict photojournalism with mobile photography apps. *New Media and Society, 16*(8), 1233–1248.

Amason, A. (1996). Distinguishing effects of functional and dysfunctional conflict on strategic decision making: Resolving a paradox for top management teams. *Academy of Management Journal, 39*(1), 123–148.

Amason, A., & Sapienza, H. (1997). The effects of top management team size and interaction norms on cognitive and affective conflict. *Journal of Management, 23*(4), 496–516.

American Psychiatric Association. (1980). *Diagnostic and statistical manual of mental disorders* (3rd ed.). Washington, DC: Author.

Anderson, C. J. (2003). The psychology of doing nothing: Forms of decision avoidance result from reason and emotion. *Psychological Bulletin, 129*, 139–167.

Anderson, R. A., & Cui, H. (2009). Intention, action planning, and decision making in parietal–frontal circuits. *Neuron, 63*(5), 568–583.

Angrilli, A., Mauri, A., Palomba, D., Flor, H., Birbaumer, N., Sartori G., et al. (1996). Startle reflex and emotion modulation impairment after a right amygdala lesion. *Brain, 119*, 1991–2000.

Anthony A., Jr. (2008). Cognitive load theory and the role of learner experience: An abbreviated review for educational practitioners. *AACE Journal, 16*(4), 425–439.

Arkes, H. R., & Ayton, P. (1999). The sunk cost and Concorde effects: Are humans less rational than lower animals? *Psychological Bulletin, 125*(5), 591–600.

Ashcraft, M. H., & Kirk, E. P. (2001). The relationship among working memory, math anxiety, and performance. *Journal of Experimental Psychology: General, 130*, 224–237.

Atkinson, R. (2005). *In the company of soldiers: A chronicle of combat.* New York, NY: Holt.

Baber, C., Fulthorpe, C., & Houghton, R. J. (2010). Supporting naturalistic decision-making through location-based photography: a study of simulated military reconnaissance. *International Journal of Human-Computer Interaction, 26*(2-3), 147–172.

Bandura, A. (1997). *Self-efficacy: The exercise of control.* New York, NY: Freeman.

Bardi, A., & Schwartz, S. H. (2003). Values and behavior, strength and structure of relations. *Personality and Social Psychology Bulletin, 10*, 1207–1220.

Baron, J., & Spranca, M. (1997). Protected values. *Organizational Behavior and Human Decision Processes, 70*, 1–16.

Baron, R. A. (1991). Positive effects of conflict: A cognitive perspective. *Employee Responsibilities and Rights Journal, 4*(1), 25–36.

Barrett, E., & Martin, P. (2014). *Extreme: Why some people thrive at the limits.* Oxford, UK: Oxford University Press.

Bartone, P. T. (1999). Hardiness protects against war-related stress in Army Reserve Forces. *Consulting Psychology Journal: Practice and Research, 51*(2), 72–82.

Baucus, M. S., & Beck-Dudley, C. L. (2005). Designing ethical organizations: Avoiding the long-term negative effects of rewards and punishments. *Journal of Business Ethics, 56*, 355–376.

Beeler, J. D., & Hunton, J. E. (1997). The influence of compensation method and disclosure level on information search strategy and escalation of commitment. *Journal of Behavioral Decision Making, 10*, 77–91.

Beike, D. R., Markman, K. D., & Karadogan, F. (2009). What we regret most are lost opportunities: A theory of regret intensity. *Personality and Social Psychology Bulletin, 35*, 385–397.

Beilock, S. L., Carr, T. H., MacMahon, C., & Starkes, J. L. (2002). When paying attention becomes counterproductive: Impact of divided versus skill-focused attention on novice and experienced performance of sensorimotor skills. *Journal of Experimental Psychology: Applied, 8*, 6–16.

Belenky, G. (1997). *Sleep, sleep deprivation, and human performance in continuous operations.* Paper presented at the Joint Services Conference on Professional Ethics— JSCOPE '97. Retrieved July 30, 2010, from http://isme.tamu.edu/JSCOPE97/Belenky97/Belenky97.htm

Ben-Shalon, U., Klar, Y., & Benbenisty, Y (2012). Characteristics of sensemaking in combat. In J. H. Laurence & M. D. Matthews (Eds.), *The Oxford handbook of military psychology* (pp. 218–231). New York, NY: Oxford University Press.

Ben-Shalom, U., Lehrer, Z., & Ben-Ari, E. (2005). Cohesion during military operations: A field study on combat units in the Al-Aqsa Intifada. *Armed Forces & Society, 32*(1), 63–79.

Bennet, G. (2011). Shakespeare and post-traumatic stress disorder-extra. *British Journal of Psychiatry, 198*(4), 255.

Bessen, R. J., Belcher, V. W., & Franklin, R. J. (1987). Rucksack paralysis with and without rucksack frames. *Military Medicine, 152*(7), 372–375.

Bonnar, J. (1824). Observation on the employment of blood-letting in the cure of disease. *Edinburgh Medical and Surgical Journal, 22*, 31–47.

Boon, S. D., & Holmes, J. G. (1991). The dynamics of interpersonal trust: Resolving uncertainty in the face of risk. In R. A. Hinde & J. Groebel (Eds.), *Cooperation and prosocial behaviour* (pp. 190–211). Cambridge, UK: Cambridge University Press.

Bordin, J. (2011). *A crisis of trust and cultural incompatibility: A Red Team study of mutual perceptions of Afghan National Security Force personnel and U.S. soldiers in understanding and mitigating the phenomena of ANSF-committed fratricide*. N2KL Red Team Political and Military Behavioral Scientist Report.

Boulding, K. (1963). *Conflict and defense*. New York, NY: Harper & Row.

Bourgeois, L. J. (1985). Strategic goals, perceived uncertainty, and economic performance in volatile environments. *Academy of Management Journal, 28*(3), 548–573.

Boyd, J. R. (1996). *The essence of winning and losing*. Retrieved from https://fasttransients.files.wordpress.com/2010/03/essence_of_winning_losing.pdf

Bratman, M. E. (1987). *Intention, plans, and practical reason*. Cambridge, MA: Harvard University Press.

Braybrooke, D., & Lindlom, C. (1963). *A strategy of decision: Policy evaluation as a social process*. New York, NY: Free Press.

Brewin, C. R., Andrews, B., & Valentine, J. D. (2000). Meta-analysis of risk factors for posttraumatic stress disorder in trauma-exposed adults. *Journal of Consulting and Clinical Psychology, 68*, 748–766.

Bryant, D. J. (2002). Making naturalistic decision making "fast and frugal." In *Proceedings of the 7th International Command and Control Research Program*. Toronto, Ontario, Canada: Defence Research Development Canada. http://www.dodccrp.org

Buckworth, J., & Dishman, R. K. (2002). *Exercise psychology*. Champaign, IL: Human Kinetics.

Budescu, D. V., & Rantilla, A. K. (2000). Confidence in aggregation of expert opinions. *Acta Psychologica, 104*, 371–398.

Burwell, D. W. Maj (2001). *Logical Evolution of the Military Decision Making Process*. Leavenworth: School of Advanced Military Studies, US Army Command and General Staff College.

Call, S. (2007). *Danger close: Tactical air controllers in Afghanistan and Iraq*. Williams-Ford TX: Texas A&M University Press.

Cannon-Bowers, J. A., & Bell, H. H. (1997). Training decision makers for complex environments: Implications of the naturalistic decision making perspective. In C. E. Zsambok & G. A. Klein (Eds.), *Naturalistic decision making* (pp. 99–110). Mahwah, NJ: Erlbaum.

Cassel, R. (1993). Building trust in Air Force leadership. *Psychology: A Journal of Human Behavior, 30*(3), 4–15.

Catignani, S. (2012). "Getting COIN" at the tactical level in Afghanistan: Reassessing counter-insurgency adaptation in the British Army. *Journal of Strategic Studies, 35*, 513–539.

Chater, N., & Loewenstein, G. (2016). The under-appreciated drive for sense-making. *Journal of Economic Behavior & Organization, 126*(2), 137–154.

Chatman, J. (1989). Improving interactional organizational research: A model of person–organization fit. *Academy of Management Review, 14*(3), 133–138.

Chatman, J. (1991). Matching people and organizations: Selection and socialization in public accounting firms. *Administrative Science Quarterly, 36*(3), 459–484.

Chee, M. W. L., & Choo, W. C. (2004). Functional imaging of working memory after 24 hr of total sleep deprivation. *Journal of Neuroscience, 24* (19), 4560–4567.

Clausewitz, C. (1984). *On war* (rev. ed.; M. Howard & P. Paret, trans.). Princeton, NJ: Princeton University Press.

Clinton, H. (2014). *Hard choices*. London, UK: Simon & Schuster.

Cohen, M. S., Freeman, J. T., & Wolf, S. (1996). Meta-recognition in time stressed decision making: Recognizing, critiquing, and correcting. *Human Factors, 38*(2), 206–219.

Cohen, M. S., & Lipshitz, R. (2011). *Three roads to commitment: A trimodal theory of decision making*. Unpublished manuscript.

Coker, A. L., Smith, P. H., & Fadden, M. K. (2005). Intimate partner violence and disabilities among women attending family practice clinics. *Journal of Womens Health, 14*(9), 829–838.

Costa, A. C., Roe, R. A., & Taillieu, T. (2001). Trust within teams: The relation with performance effectiveness. *European Journal of Work and Organisational Psychology, 10*(3), 225–244.

Costa, P. T., & McCrae, R. R. (1992). *Revised NEO personality inventory & NEO Five Factor inventory: Professional manual*. Odessa, FL: Psychological Assessment Resources.

Coyne, J. C., & Lazarus, R. S. (1980). Cognitive style, stress perception, and coping. In I. L. Kutash & L. B. Schesinger (Eds.), *Handbook on stress and anxiety: Contemporary knowledge, theory and treatment* (pp. 144–158). San Francisco, CA: Jossey-Bass.

Crandall, B., Klein, G., & Hoffman, R. R. (2006). *Working minds: A practitioner's guide to cognitive task analysis*. Cambridge, MA: MIT Press.

Currier, J. M., Holland, J. M., & Malott, J. (2014). Moral injury, meaning making, and mental health in returning veterans. *Journal of Clinical Psychology, 00*(0), 1–12. doi:10.1002/jclp.22134.

David, S. (1997). *Military blunders: The how and why of military failure*. London, UK: Robinson.

De Dreu, C. K. W., & Weingart, L. R. (2003). Task versus relationship conflict, team performance and team member satisfaction: A meta-analysis. *Journal of Applied Psychology, 88*(4), 741–749.

de Kwaadsteniet, E. W., van Dijk, E., Wit, A., De Cremer, D., & de Rooij, M. (2007). Justifying decisions in social dilemmas: Justifying pressured and tacit coordination under environmental uncertainty. *Personality & Social Psychology Bulletin, 33*(12), 1648–1660.

de Wit, F. R. C., Greer, L. L., & Jehn, K. A. (2012). The paradox of intragroup conflict: A meta-analysis. *Journal of Applied Psychology, 97*(2), 360–390.

DeChurch, L. A., & Marks, M. A. (2001). Maximizing the benefits of task conflict: The role of conflict management. *International Journal of Conflict Management, 12*(1), 4–22.

Department of the Army. (2011). *Commander and staff officer guide*. Washington, DC: Author.

Deputy Under Secretary of the Army Knowledge Center. Aug 25, 2009. [accessed May 21, 2009]. Available at: http://www.army.mil/armyBTKC/focus/sa/about.htm.

Dhar, R. (1997). Context and task effects on choice deferral. *Marketing Letters, 8*(1), 119–130.

Dhar, R., & Nowlis, S. (1999). The effect of time pressure on consumer choice deferral. *Journal of Consumer Research, 25,* 369–384.

Dhar, R., & Simonson, I. (2003). The effect of forced choice on choice. *Journal of Marketing Research, 40,* 146–160.

Diertrich, A. (2006). Transient hypofrontality as a mechanism for the psychological effects of exercise. *Psychiatric Research, 145*(1), 79–83.

Doyle, J., & Thomason, R. (1999). Background to qualitative decision theory. *Artificial Intelligence Magazine, 20*(2), 55–68.

Duc, C., Hanselmann, M., Boesiger, P., & Tanner, C. (2013). Sacred values: Trade-off type matters. *Journal of Neuroscience, Psychology, and Economics, 6*(4), 252–263.

Eddy, M., Hasselquist, L., Giles, G., Hayes, J., Howe, J., Rourke, J., ... Mahoney, C. R. (2015). The effects of load carriage and physical fatigue on cognitive performance. *PLoS One, 10*(7), e0130817.

Eisenhardt, K. M., & Schoonhoven, C. B. (1990). Organizational growth: Linking founding team, strategy, environment, and growth among US semiconductor ventures, 1978–1988. *Administrative Science Quarterly, 35*(3), 504–529.

Elliot, A. J. (2006). The hierarchical model of approach–avoidance motivation. *Motivations and Emotion, 30,* 111–116.

Elliot, A. J., Eder, A. B., & Harmon-Jones, E. (2013). Approach–avoidance motivation and emotion: Convergence and divergence. *Emotion Review, 5*(3), 308–311.

Elliott, T. (2005). *Expert decision-making in naturalistic environments: A summary of research.* Edinburgh, South Australia: Australian Government, Department of Defence, Defence Science and Technology Organisation.

Endsley, M. R. (1988). Design and evaluation for situation awareness enhancement. In *Proceedings of the Human Factors Society 32nd Annual Meeting* (pp. 97–101). Santa Monica, CA: Human Factors and Ergonomics Society.

Endsley, M. R. (1993). *Situational awareness in dynamic human decision-making: Theory.* Paper presented at the First International Conference on Situational Awareness in Complex Systems, Orlando, FL.

Endsley, M. R. (1995a): Toward a theory of situation awareness in dynamic systems. *Human Factors, 37*(1), 32–64.

Endsley, M. R. (1995b). Measurement of situation awareness in dynamic systems. *Human Factors, 37,* 65–84.

Endsley, M. R. (2000). Theoretical underpinnings of situation awareness: A critical review. In M. R. Endsley & F. J. Garland (Eds.), *Situation awareness analysis and measurement* (pp. 3–32). Mahwah, NJ: Erlbaum.

Endsley, M. R., & Garland, D. G. (Eds.). (2000). *Situation awareness analysis and measurement.* Mahwah, NJ: Erlbaum.

Epstein, S., & Fenz, W. D. (1965). Steepness of approach and avoidance gradients in humans as a function of experience. *Journal of Experimental Psychology, 70,* 1–12.

Eriksen, J. W. (2010). Should soldiers think before they shoot? *Journal of Military Ethics, 9*(3), 195–218.

Eyre, M., & Alison, L. (2010). Investigative Decision Making. In J. Brown & E. Campbell (Eds.), *The Cambridge Handbook of Investigative Psychology* (pp. 73–80). Cambridge: Cambridge University Press.

Eyre, M., Alison, L., Crego, J., & Mclean C. (2008). Decision inertia: The impact of organisations on critical incident decision-making. In L. J. Alison & J. Crego (Eds.),

Policing critical incidents: Leadership and critical incident management. Devon, UK: Willan.

Eysenck, M. W., & Keane, M. T. (1990). *Cognitive psychology: A student's handbook.* Hillsdale, NJ: Erlbaum.

Fallesen, J. J. (1993). *Overview of Army tactical planning performance research* (Technical Report 984, pp. 42–44), Fort Leavenworth, KS: US Army Research Institute for the Behavioral and Social Sciences.

Faris, J. H. (1975). The impact of basic combat training: The role of the drill sergeant in the all-volunteer army. *Armed Forces & Society, 2*(1), 115–127.

Farnsworth, J., Drescher, K., Nieuwsma, J., Walser, R., & Currier, J. (2014). The role of moral emotions in military trauma: Implications for the study and treatment of moral injury. *Review of General Psychology, 18*, 249–262. http://dx.doi.org/10.1037/gpr0000018

Farrell, K. (2011). *Berserk style in American culture.* New York, NY: Palgrave Macmillan.

Farrell, T., Osinga, F., & Russell, J. A. (2013). *Military adaptation in Afghanistan.* Stanford, CA: Stanford University Press.

Feldman, M. L., & Spratt, M. F. (1998). *Five frogs on a log.* New York: HarperCollins.

Fellows, L. K. (2004). The cognitive neuroscience of human decision making: A review and conceptual framework. *Behavioural Cognitive Neuroscience Review, 3*, 159–172.

Fick, N. (2006). *One bullet away: The making of a Marine officer.* New York, NY: Houghton Mifflin.

Filkins, D. (2014, November 19). The long road home: *Redeployment*, by Phil Klay. *New York Times*. Retrieved from https://www.nytimes.com/2014/03/09/books/review/redeployment-by-phil-klay.html

Fischer, G., Jia, J., & Luce, M. (2000). Attribute conflict and preference uncertainty: The RandMAU model. *Management Science, 46*, 669–684.

Fischer, G., Luce, M., & Jia, J. (2000). Attribute conflict and preference uncertainty: Effects on judgment time and error. *Management Science, 46*, 88–103.

Fischhoff, B. (1975). Hindsight is not equal to foresight: The effect of outcome knowledge on judgment under uncertainty. *Journal of Experimental Psychology: Human Perception and Performance, 1*(3), 288–299.

Fiske, A. P., & Tetlock, P. E. (1997). Taboo tradeoffs: Reactions to transactions that transgress spheres of justice. *Political Psychology, 18*(2), 255–297.

Fiske, S. T., & Taylor, S. E. (1991). *Social cognition* (2nd ed.). New York, NY: McGraw-Hill.

FM 21-18 (1990). Field Manual No. 21-18, Headquarters Department of the Army.

Folkman, S., Lazarus, R. S., Dunkel-Schetter, C., DeLongis, A., & Gruen, R. J. (1986). Dynamics of a stressful encounter: Cognitive appraisal, coping, and encounter outcomes. *Journal of Personality and Social Psychology, 50*(5), 992–1003.

Franke, V. (2011). Decision-making under uncertainty: Using case studies for teaching strategy in complex environments. *Journal of Military and Strategic Studies, 13*(2), 1–21.

Friedman, B. (2007). *The war I always wanted: The illusion of glory and the reality of war.* St. Paul, MN: Zenith Press.

Friedman, M. J., Schnurr, P. P., & McDonagh-Coyle, A. (1994). Post-traumatic stress disorder in the military veteran. *Psychiatric Clinics of North America, 17*(2), 265–278.

Funder, D. C. (1987). Errors and mistakes: Evaluating the accuracy of social judgment. *Psychological Bulletin, 101*, 75–90.

Gerras, S. J. (2002). *The Army as a learning organization* (Strategy Research Report). Carlisle Barracks, PA: Army War College.

Gigerenzer, G., & Goldstein, D. G. (1996). Reasoning the Fast and Frugal Way: Models of Bounded Rationality. *Psychological Review, 103*(4), 650–669.

Gigerenzer, G., & Todd, P. M. (1999). *Simple heuristics that make us smart*. New York, NY: Oxford University Press.

Gilbert, D. T., Brown, R. P., Pinel, E. C., & Wilson, T. D. (2000). The illusion of external agency. *Journal of Personality and Social Psychology, 79*, 690–700.

Gilbert, D. T., Driver-Linn, E., & Wilson, T. D. (2002). The trouble with Vronsky: Impact bias in the forecasting of future affective states. In L. Feldman-Barrett & P. Salovey (Eds.), *The wisdom of feeling* (pp. 114–143). New York, NY: Guilford.

Gilbert, D. T., Pinel, E. C., Wilson, T. D., Blumberg, S. J., & Wheatley, T. P. (1998). Immune neglect: A source of durability bias in affective forecasting. *Journal of Personality and Social Psychology, 75*, 617–638.

Gilbert, D. T., & Wilson, T. D. (2000). Miswanting: Some problems in the forecasting of future affective states. In J. Forgas (Ed.), *Feeling and thinking: The role of affect in social cognition* (pp. 178–197). Cambridge, UK: Cambridge University Press.

Glasher, J., et al. (2012). Lesion mapping of cognitive control and value-based decision making in the prefrontal cortex. *Proceedings of the National Academy of Sciences of the USA, 109*(36), 14681–14686.

Goldstein, D. G., & Gigerenzer, G. (2002). Models of ecological rationality: The recognition heuristic. *Psychological Review, 109*(1), 74–90.

Goldstein, W. M., & Hogarth, R. M. (1997). Judgment and decision research: Some historical context. In W. M. Goldstein & R. M. Hogarth (Eds.), *Research on judgment and decision making: Currents, connections, and controversies* (pp. 3–65). New York, NY: Cambridge University Press.

Golightly, D., Wilson, J. R., Lowe, E., & Sharples, S (2010). The role of situation awareness for understanding signaling and control in rail operations. *Theoretical Issues in Ergonomics Science, 11*, 84–98.

Gollwitzer, P. M., & Moskowitz, G. B. (1996). Goal effects on actions and cognition. In E. T. Higgins & A. W. Kruglanski (Eds.), *Social psychology: Handbook of basic principles* (pp. 361–399). New York, NY: Guilford.

Gonzalez, C., Dana, J., Koshino, H., & Just, M. (2005). The framing effect and risky decisions: Examining cognitive functions with fMRI. *Journal of Economic Psychology, 26*(1), 1–20.

Gore, J., Banks, J., Millward, L., & Kyriakidou, O. (2006). Naturalistic decision making and organizations: Reviewing pragmatic science. *Organization Studies, 27*(7), 925–942.

Gore, J., Flin, R., Stanton, N., & Wong, B. L. W. (2015). Applications for naturalistic decision-making, *Journal of Occupational and Organizational Psychology, 88* (2), 223–230.

Gow, J. (2013). *War and war crimes*. New York, NY: Columbia University Press.

Gray, J. A. (1990). Brain systems that mediate both emotion and cognition. *Cognition and Emotion, 4*(3), 269–288.

Greeley, P. J. (1995). Energizing boards, commissions, task forces, and volunteer groups. *Economic Development Review, 13*(3), 24–27.

Green, D. M., & Swets, J. A. (1966). *Signal detection theory and psychophysics* (A reprint, with corrections of the original 1966 ed.). Huntington, NY: Robert E. Krieger Publishing Co.

Greene, J. D. (2014). The cognitive neuroscience of moral judgment and decision making. In M. S. Gazzaniga (Ed.), *The cognitive neurosciences V* (pp. 1013–1023). Cambridge, MA: MIT Press.

Grossman, D. (1996). *On killing: The psychological cost of learning to kill in war and society.* New York, NY: Back Bay Books.

Guetzkow, H., & Gyr, J. (1954). An analysis of conflidt in decision-making groups. *Human Relations, 7*(3), 367–382.

Gurvits, T. V., Shenton, M. E., Hokama, H., Ohta, H., Lasko, N. B., Gilbertson, M. W., Orr, S. P., Kikinis, R., Jolesz, F. A., McCarley, R. W., & Pitman, R. K. (1996). A magnetic resonance imaging study of hippocampal volume in chronic, combat related, posttraumatic stress disorder. *Biological Psychiatry, 40,* 1091–1099.

Hall, A. T., Bowen, M. G., Ferris, G. R., Royle, M. T., & Fitzgibbons, D. E. (2007). The accountability lens: A new way to view management issues. *Business Horizons, 50,* 405–413.

Hannah, S., Campbell, D., & Matthews, M. (2010). Advancing a research agenda for leadership in dangerous contexts. *Military Psychology, 22*(1), 157–189.

Hanselmann, M., & Tanner, C. (2008). Taboos and conflicts in decision making: Sacred values, decision difficulty, and emotions. *Journal of Judgement and Decision Making, 3*(1), 51–63.

Harrison, Y., & Horne, J. A. (2000). The impact of sleep deprivation on decision making: A review. *Journal of Experimental Psychology: Applied, 6,* 236–249.

Hart, P. 't. (1990). *Groupthink in government: A study of small groups and policy failure.* Baltimore, MD: Johns Hopkins University Press.

Hartle, A. E. (1989). *Moral issues in military decision making.* Lawrence, KS: University Press of Kansas.

Hawkins, S. A., & Hastie, R. (1990). Hindsight: Biased judgments of past events after the outcomes are known. *Psychological Bulletin, 107,* 311–327.

Hensley, L., & Varela, E. (2008). PTSD symptoms and somatic complaints following Hurricane Katrina: The roles of trait anxiety and anxiety sensitivity. *Journal of Clinical Child & Adolescent Psychology, 37*(3), 542–552.

Hittle, J. D. (1975). *The military staff: Its history and development.* Westport, CT: Greenwood.

Hockey, G. R. J. (1997). Compensatory control in the regulation of human performance under stress and high workload: A cognitive-energetical framework. *Biological Psychology, 45*(3), 73–93.

House, A., Power, N., & Alison, L. (2014). A systematic review of the potential hurdles of interoperability to the emergency services in major incidents: Recommendations for solutions and alternatives. *Cognition, Technology & Work, 16,* 319–335.

Huemer, L., von Krogh, G., & Roos, J. (2000). Knowledge and the concept of trust. In G. von Krogh & D. Kleine (Eds.), *Knowing in firms: Understanding, managing and measuring knowledge* (pp. 122–145). London, UK: Sage.

Hunter, S.T., Cushenbery, L., & Friedrich, T. (2012). Hiring an innovative workforce: A necessary yet uniquely challenging endeavor. *Human Resource Management Review, 22*(4), 303–322.

Inman, J. J., & Zeelenberg, M. (1998, February). What might be: The role of potential regret in consumer choice. In L. J. Abendroth (Chair), *Regret me not: An examination of regret in pre- and post-purchase evaluations.* Symposium conducted at the Society for Consumer Psychology 1998 Winter Conference, Austin, TX.

Ivy, L. (1995). *A study in leadership: The 761st Tank Battalion and the 92nd Division in World War II.* Fort Leavenworth, KS: United States Army Command and General Staff College.

Jackson, J. J., Thoemmes, F., Jonkmann, K., Lüdtke, O., & Trautwein, U. (2012). Military training and personality trait development: Does the military make the man or does the man make the military. *Psychological Science, 23*(3), 270–277. http://dx.doi.org/10.1177/0956797611423545

Janis, I. L. (1982). *Groupthink* (2nd ed.). Boston, MA: Houghton Mifflin.

Janis, I. L., & Mann, L. (1979). *Decision making: A psychological analysis of conflict, choice and commitment.* New York, NY: Free Press.

Jeffrey, R. C. (1983). *The logic of decision.* Chicago, IL: University of Chicago Press.

Jehn, K. (1994). Enhancing effectiveness: An investigation of advantages and disadvantages of value-based intragroup conflict. *International Journal of Conflict Management, 5*(3), 223–238.

Jehn, K. (1995). A multimethod examination of the benefits and detriments of intragroup conflict. *Administrative Science Quarterly, 40*(2), 256–282.

Jehn, K. (1997). A qualitative analysis of conflict types and dimensions in organizational groups. *Administrative Science Quarterly, 42*(3), 530–557.

Jehn, K., Chadwick, C., & Thatcher, S. M. B. (1997). To agree or not to agree? The effect of value congruence, individual demographic dissimilarity and conflict on workgroup outcomes. *International Journal of Conflict Management, 8*(4), 287–305.

Jehn, K. A., & Mannix, E. A. (2001). The dynamic nature of conflict: A longitudinal study of intragroup conflict and group performance. *Academy of Management Journal, 44*(2), 238–251.

Jones, D. G., & Endsley, M. R. (1996). Sources of situation awareness errors in aviation. *Aviation, Space, and Environmental Medicine, 67*(6), 507–512.

Kaempf, G. L., Klein, G. A., Thordsen, M. L., & Wolf, S. (1996). Decision making in complex Naval command-and-control environments. *Human Factors, 38*(2), 220–231.

Kahl, C. H. (2007). In the crossfire or the crosshairs? Norms, civilian casualties, and US conduct in Iraq. *International Security, 32*(1), 7–46.

Kahneman, D. (1994). New challenges to the rationality assumption. *Journal of Institutional and Theoretical Economics, 150,* 18–36.

Kahneman, D. (2011). *Thinking, fast and slow.* New York, NY: Farrar, Straus & Giroux.

Kahneman, D., Slovic, P., & Tversky, A. (Eds.). (1982). *Judgment under uncertainty: Heuristics and biases.* New York, NY: Cambridge University Press.

Kahneman, D., & Tversky, A. (1982). The psychology of preferences. *Scientific American, 246,* 160–173.

Kelley, H. H. (1973). The process of causal attribution. *American Psychologist, 28,* 107–128.

Kennedy, L. (2008). Securing vision: Photography and the US foreign policy. *Media, Culture & Society, 30*(3): 279–294.

Kessler, R. C., Berglund, P., Delmer, O., Jin, R., Merikangas, K. R., & Walters, E. E. (2005). Lifetime prevalence and age-of-onset distributions of DSM-IV disorders in the National Comorbidity Survey Replication. *Archives of General Psychiatry, 62*(6), 593–602.

King, A. (2006). The world of command: Communication and cohesion in the military. *Armed Forces and Society, 32*(4), 493–512.

Klee, S., & Renner, K.-H. (2016). Beyond pride and prejudices: An empirical investigation of German armed forces soldiers' personality traits. *Personality and Individual Differences, 88,* 261–266.

Klehe, U., Anderson, N., & Hoefnagels, E. A. (2007). Social facilitation and inhibition during maximum versus typical performance situations. *Human Performance, 20*(3), 223–239.

Klein, G. (1989, May). Strategies of decision making. *Military Review, 56*, 56–64.
Klein, G. (1993). A recognition-primed decision (RPD) model of rapid decision making. In G. Klein, J. Orasanu, R. Calderwood, & C. E. Zsambok (Eds.), *Decision making in action: Models and methods* (pp. 138–147). Norwood, CT: Ablex.
Klein, G. (1997). Naturalistic Decision Making: Where Are We Going? In C. E. Zsambok & S. G. Klein (Eds.), *Naturalistic Decision Making* (pp. 383–399). Mahwah, NJ: L. Erlbaum Associates.
Klein, G. (1998). *Sources of power: How people make decisions*. Cambridge, MA: MIT Press.
Klein, G. (2008). Naturalistic decision making. *Human Factors, 50*(3), 456–460.
Klein, G. (2011). *Streetlights and shadows—Searching for the keys to adaptive decision making*. Cambridge, MA: MIT Press.
Klein, G. A., Calderwood, R., & Clinton-Cirocco, A. (1986). Rapid decision making on the fireground. In *Proceedings of the Human Factors and Ergonomics Society 30th Annual Meeting* (Vol. 1, pp. 576–580). Norwood, NJ: Ablex.
Klein, G., Calderwood, R., & Clinton-Cirocco, A. (2010). Rapid decision making on the fireground: The original study plus a postscript. *Journal of Cognitive Engineering and Decision Making, 4*(3), 186–209.
Klein, G., & Hoffman, R. R. (1993). Perceptual–cognitive aspects of expertise. In M. Rabinowiz (Ed.), *Cognitive science foundations of instruction* (pp. 203–226). Hillsdale, NJ: Erlbaum.
Klein, G., Moon, B., & Hoffman, R. F. (2006). Making sense of sensemaking: II. A macrocognitive model. *IEEE Intelligent Systems, 21*, 88–92.
Klein, G., Orasanu, J., Calderwood, R., & Zsambok, C. E. (Eds.). (1993). Decision making in action: Models and methods. Norwood, CT: Ablex.
Klein, G., Phillips, J. K., Klinger, D. W., & McCloskey, M. J. (1998). *The Urban Warrior Experiment: Observations and recommendation for ECOC functioning*. Fairborn, OH: Klein Associates.
Klein, G. A., Calderwood, R., & Macgregor, D. (1989). Critical decision method for eliciting knowledge. *IEEE Transactions on Systems, Man, and Cybernetics, 19*, 462–472.
Klein, G. A., & Crandall, B. W. (1995). The role of mental simulation in naturalistic decision making. In P. Hancock, J. Flach, J. Caird, & K. Vincente (Eds.), *Local applications of the ecological approach to human–machine systems* (Vol. 2, pp. 324–358). Hillsdale, NJ: Erlbaum.
Kobus, D. A., Brown, C. M., Wu, L., Robusto, K., & Bartlett, J. (2011). *Cognitive performance and physiological changes under heavy load carriage*. San Diego, CA: Pacific Science and Engineering Group, Naval Health Research Centre.
Kolditz, T. A. (2006). Research in "in extremis" settings. *Armed Forces & Society, 32*(4), 655–658.
Koren, D., Norman, D., Cohen, A., Berman, J., & Klein, E. M. (2005). Increased PTSD risk with combat-related injury: a matched comparison study of injured and uninjured soldiers experiencing the same combat events. *American Journal of Psychiatry, 162*, 276–282.
Kramer, R. M. (1999). Trust and distrust in organizations: Emerging perspectives, enduring questions. *Annual Review of Psychology, 50*, 569–598.
Kristof, A. L. (1996). Person–organization fit: An integrative review of its conceptualizations, measurement, and implications. *Personnel Psychology, 49*(1), 1–49.

Kruglanski, A. W., & Stroebe, W. (2005). The influence of beliefs and goals on attitudes. In D. Albarracin, B. T. Johnson, & M. T. Zanna (Eds.), *Handbook of attitudes* (pp. 323–368). Mahwah, NJ: Erlbaum.

Kulka, R. A., Cchlenger, W. E., Fairbank, J. A., Hough, R. L., Jordan, B. K., Marmar, C. R., & Weiss, D. S. (1990). *Trauma and the Vietnam war generation*. New York: Brunner/Mazel, Inc.

Larsen, R. P. (2001). Decision making by military students under severe stress. *Military Psychology, 13*(2), 89–98.

Lazarus, R. S. (1981). The stress and coping paradigm, In C. Eisdorfer, D Cohen, A. Kleinman, & P. Maxi (Eds.), *Models for clinical psychopathology* (pp. 177–214). New York, NY: Spectrum.

Lazarus, R. S., & Folkman, S. (1984). *Stress, appraisal, and coping*. New York, NY: Springer.

Lester, D., & Brockopp, G. W. (Eds). (1973). *Crisis intervention and counseling by telephone*. Springfield, IL: Charles C. Thomas.

Lewin, K. (1951). *Field theory in social science: Selected theoretical papers*. New York, NY: Harper & Row.

Liberzon, I., Abelson, J. L., Flagel, S. B., Raz, J., & Young, E. A. (1999). Neuroendocrine and psychophysiologic responses in PTSD: A symptom provocation study. *Neuropsychopharmacology, 21*, 40–50.

Liedtka, J. (1989). Value congruence: The interplay of individual and organizational value systems. *Journal of Business Ethics, 8*, 805–815.

Lim, B. C., & Ployhart, R. E. (2004). Transformational leadership: Relations to the five-factor model and team performance in typical and maximum contexts. *Journal of Applied Psychology, 89*(4), 610.

Lipshitz, R. (1993). Converging themes in the study of decision making in realistic settings. In G. A. Klein, J. Orasanu, R. Calderwood, & C. E. Zsambok (Eds.), *Decision making in action: Models and methods* (pp. 103–137). Norwood, CT: Ablex.

Lipshitz, R., & Bar-Ilan, O. (1996). How problems are solved: Reconsidering the phase theorem. *Organizational Behaviour and Human Decision Processes, 65*(1), 48–60.

Lipshitz, R., Klein, G., Orasanu, J., & Salas, E. (2001). Focus article: Taking stock of naturalistic decision making. *Journal of Behavioral Decision Making, 14*, 331–352.

Lipshitz, R., & Strauss, O. (1997). Coping with uncertainty: A naturalistic decision-making analysis. *Organizational Behavior and Human Decision Processes, 69*(2), 149–163.

Litman, R. E. (1971). Suicide prevention: Evaluating effectiveness. *Life-Threatening Behavior, 1*, 155–162.

Litz, B. T., Stein, N., Delaney, E., Lebowitz, L., Nash, W. P., Silva, C., & Maguen, S. (2009). Moral injury and moral repair in war veterans: A preliminary model and intervention strategy. *Clinical Psychology Review, 29*, 695–706.

Loewenstein, G., & Frederick, S. (1997). Predicting reactions to environmental change. In M. H. Bazerman, D. M. Messick, A. E. Tenbrusel, & K. A. Wade-Benzoni (Eds.), *Environment, ethics, and behavior* (pp. 52–72). San Francisco, CA: New Lexington Press.

Loewenstein, G., & Prelec, D. (1993). Preferences for sequences of outcomes. *Psychological Review, 100*, 91–108.

Loewenstein, G. F., & Schkade, D. (1999). Wouldn't it be nice? Predicting future feelings. In D. Kahneman, E. Diener, & N. Schwartz (Eds.), *Well-being: The foundations of hedonic psychology* (pp. 85–105). New York, NY: Russell Sage Foundation.

Luce, M., Payne, J., & Bettman, J. (2001). The impact of emotional tradeoff difficulty on decision behavior. In E. U. Weber, J. Baron, & G. Loomes (Eds.), *Conflict and tradeoffs in decision making* (pp. 86–109). New York, NY: Cambridge University Press.

Luhmann, N. (1988). Familiarity, confidence, trust: Problems and alternatives. In D. Gambetta (Ed.). *Trust: Making and breaking cooperative relations* (pp. 94–108). New York, NY: Basil Blackwell.

Luttrell, M. (2007). *Lone survivor: The eyewitness account of Operation Redwing and the lost heroes of SEAL team 10*. New York, NY: Little, Brown.

Maddi, S. R. (2007). Hardiness: The courage to grow from stresses. *Journal of Positive Psychology, 1*(3), 160–168.

Mahoney, C. R., Hirsch, E., Hasselquist, L., Lesher, L. L., & Lieberman, H. R. (2007). The effects of movement and physical exertion on soldier vigilance. *Aviation, Space and Environmental Medicine, 78*(5), 51–57.

Majchrzak, A., Jarvenpaa, S. L., & Hollingshead, A. B. (2007). Coordinating expertise among emergent groups responding to disasters. *Organization Science, 18*(1), 147–161.

Malinowski, B. (1944). *A scientific theory of culture*. Chapel Hill, NC: University of North Carolina Press.

Mamhidir, A. G., Kihlgren, M., & Sorlie, V. (2007). Ethical challenges related to elder care: High-level decision-makers' experiences. *BMC Medical Ethics, 8*(3), 1–10.

Manson, M. (2016). *The subtle art of not giving a f*ck*. New York, NY: HarperCollins.

March, J. G., & Olsen, J. P. (Eds.). (1976). *Ambiguity and choice in organizations*. Bergen, Norway: Universitetsforlaget.

Marshall, S. L. A. (1947). *Men against fire*. New York, NY: Morrow.

Matthews, M. (2013). *Head Strong: How psychology is revolutionizing war*. New York, NY: Oxford University Press.

Mayer, R. C., Davis, J. H., & Schoorman, F. D. (1995). An integrative model of organizational trust. *Academy of Management Review, 20*(3), 709–734.

McChrystal, S. (General). (2009). COMISAF/CDR USFOR-A, "Tactical directive," July 6, 2009 (unclassified version).

McKay, J. (1991). Assessing aspects of object relations associated with immune function: Development of the affiliative trust–mistrust coding system. *Psychological Assessment, 3*(4), 641–647.

McNally, R. J. (1989). Is anxiety sensitivity distinguishable from trait anxiety? Reply to Lilienfeld, Jacob, and Turner. *Journal of Abnormal Psychology, 98*(2), 193–194.

McNally, R. J. (2003). Progress and controversy in the study of posttraumatic stress disorder. *Annual Review of Psychology, 54*, 229.

Mercier, H., & Sperber, D. (2011). Why do humans reason? Arguments for an argumentative theory. *Behavioral and Brain Sciences, 34*, 57–74.

Meyerson, D., Weick, K. E., & Kramer, R. M. (1996). Swift trust and temporary groups. In R. M. Kramer & T. R. Tyler (Eds.), *Trust in organizations: Frontiers of theory and research* (pp. 166–195). Thousand Oaks, CA: Sage.

Michel, R. R. (1990). *Historical development of the estimate of the situation* (Research Report 1577). Fort Leavenworth, KS: US Army Research Institute for the Behavioral and Social Sciences.

Michael, T., Halligan, S. L., Clark, D. M., & Ehlers, A. (2007). Rumi-nation in posttraumatic stress disorder. *Depression and Anxiety, 24*, 307–317.

Miller, A. D., & Rottinghaus, P. J. (2014). Career indecision, meaning in life, and anxiety: An existential framework. *Journal of Career Assessment, 22*(2), 233–247.

Miller, T. E., Zsambok, C. E., & Klein, G. (1997). *Commander's cognitive demands in OOTW.* Fairborn, OH: Klein Associates.

Milosevic, I., & McCabe, R. E. (2015). *Phobias: The psychology of irrational fear.* Santa Barbara, CA: Greenwood.

Mintzberg, H., Raisinghani, D., & Théorêt, A. (1976). The structure of "unstructured" decision processes. *Administrative Science Quarterly, 21*(2), 246–275.

Moore, R. J. (Captain), et al.; US Army Research Institute of Environmental Medicine. (1992). *Changes in soldier nutritional status and immune function during the Ranger training course.* Natick, MA: US Army Research Institute of Environmental Medicine.

Morris, M. W., & Moore, P.C. (2000). The lessons we (don't) learn: Counterfactual thinking and organizational accountability after a close call. *Administrative Science Quarterly, 45,* 737–765.

Morrison, J. G., Kelly, R. T., & Hutchins, S. G. (1996). Impact of naturalistic decision support on tactical situation awareness. In *Proceedings of the Human Factors Society 40th Annual Meeting.* Santa Monica, CA: Human Factors and Ergonomics Society.

Nabi, H., Consoli, S., Chastang, J., Chiron, M., Lafont, S., & Lagarde, E. (2005). Type A Behavior Pattern, Risky Driving Behaviors, and Serious Road Traffic Accidents: a Prospective Study of the GAZEL Cohort. *American Journal of Epidemiology, 161*(9), 864–870.

Nagl, J. (2012). *Learning to eat soup with a knife: Counterinsurgency lessons from Malaya and Vietnam.* Chicago, IL: University of Chicago Press.

National Defense University, Instistute for Strategic Studies (2000). *Joint Vision 2020. America's Military—Preparing for Tomorrow.* Washibngton D.C.: National Defense University, Instistute for Strategic Studies.

Norris, F. H. (1992). Epidemiology of trauma: frequency and impact of different potentially traumatic events on different demographic groups. *Journal of Consulting and Clinical Psychology, 60*(3), 409–418.

Olson, J., Roese, N., & Zanna, M. (1996). Expectancies. In E. Higgins & A. Kruglanski (Eds.), *Social psychology: Handbook of basic principles* (pp. 211–238). London, UK: Guildford.

Oppel R. A. Jr. (2010). Tighter rules fail to stem deaths of innocent Afghans at checkpoints. New York Times. Retrieved from http://www.nytimes.com/2010/03/27/world/asia/27afghan.html

Orasanu, J., & Connolly, T. (1993). The reinvention of decision making. In G. A. Klein, J. Orasanu, R. Calderwood, & C. E. Zsambok (Eds.), *Decision making in action: Models and methods* (pp. 3–20). Norwood, NJ: Ablex.

Orasanu, J., Martin, L., & Davison, J. (1998). *Errors in aviation decision making: Bad decisions or bad luck.* Paper presented at the Fourth Conference on Naturalistic Decision Making, Warrenton, VA.

Ozer, E. J., Best, S. R., Lipsey, T. L., & Weiss, D. S. (2003). Predictors of posttraumatic stress disorder and symptoms in adults: A meta-analysis. *Psychological Bulletin, 129,* 52–71.

Parker, J. R., & Schrift, R. Y. (2011). Rejectable choice sets: How seemingly irrelevant no-choice options affect consumer decision processes. *Journal of Marketing Association, 48,* 840–854.

Pascual, R., & Henderson, S. (1997). Evidence of naturalistic decision making in military command and control. In G. K.Caroline & E. Zsambok (Eds.), *Naturalistic decision making* (pp. 217–226). Mahwah, NJ: Erlbaum.

Pascual, R. G., Blendell, C., Molloy, J. J., Catchpole, L. J., & Henderson, S. M. (2004). *An investigation of alternative command planning processes*. Defence Evaluation and Research Agency (DERA) paper prepared for the UK Ministry of Defence.

Payne, J. W. (1976). Task complexity and contingent processing in decision making: An information search and protocol analysis. *Organizational Behavior and Human Performance, 16*(2), 366–387.

Pech, R., & Slade, B. (2004). Manoeuvre theory: Business mission analysis process for high intensity conflict. *Management Decision, 42*(8), 987–1000.

PEDU. (2012). *Review of response to flooding on 27th and 28th June 2012*. Retrieved from http://www.drdni.gov.uk/pedu-review-flood-response-june-2012.pdf

Pelled, L. (1996). Demographic diversity, conflict, and work group outcomes: An intervening process theory. *Organization Science, 7*(6), 615–631.

Pennington, N., & Hastie, R. (1993). A theory of explanation-based decision making. In G. Klein, J. Orasanu, R. Calderwood, & C. E. Zsambok (Eds.), Decision making in action: Models and methods (pp. 188–201). Norwood, CT: Ablex.

Peterson, R. S., & Behfar, K. J. (2003). The dynamic relationship between performance feedback, trust and conflict in groups: A longitudinal study. *Organizational Behavior and Human Decision Processes, 92*, 102–112.

Peterson, C., & Seligman, M. E. P. (2004). *Character strengths and virtues: A handbook and classijkation*. Washington, DC: American Psychological Association & Oxford University Press.

Pfaff, M. S., Klein, G. L., Drury, J. L., Moon, S. P., Liu, Y., & Entezari, S. O. (2013). Supporting complex decision making through option awareness. *Journal of Cognitive Engineering and Decision Making, 7*(2), 155–178.

Pilcher, J. J., & Huffcutt, A. J. (1996). Effects of sleep deprivation on performance: A meta-analysis. *Sleep, 19*(4), 318–326.

Pleban, R. J., Valentine, P. J., Penetar, D. M., Redmond, D. P., & Belenky, G. L. (1990). Characterization of sleep and body composition changes during ranger training. *Military Psychology, 2*(3), 145–156.

Plous, S. (1993). *The psychology of judgment and decision making*. New York, NY: McGraw-Hill.

Posen, B. R. (1984). *The sources of military doctrine: France, Britain, and Germany between the World Wars*. Ithaca, NY: Cornell University Press.

Posner, K. (2010). *Stalking the black swan: Research and decision making in a world of extreme volatility*. New York, NY: Columbia Business School.

Power, N. (2016). *Cognition in crisis: Decision inertia and failures to take action in multi-agency emergency response command teams*. Retrieved from the University of Liverpool Repository at http://livrepository.liverpool.ac.uk/2028122/1/PowerNic_June2015_2028122.pdf.

Power, N., & Alison, L. J. (2017). Redundant deliberation about negative consequences: Decision inertia in emergency responders. *Psychology, Public Policy and Law, 23*(2), 243–258.

Price, B. (2013). Syria: A wicked problem for all. *CTC Sentinel, 6*(8), 1–4.

Protopopescu, X., Pan, H., Altemus, M., Tuescher, O., Polanecsky, M., McEwen, B., et al. (2005). Orbitofrontal cortex activity related to emotional processing changes across the menstrual cycle. *Proceedings of the National Academy of Sciences, 102*, 16060–16065.

Rahe, R. H., Ryman, D. H., & Beirsner, R. J. (1976). Serum uric acid, cholesterol, and psychological moods throughout stressful naval training. *Aviation, Space and Environmental Medicine, 47*, 883–888.

Randel, J. M., Pugh, L., & Reed, S. K. (1996). Differences in expert and novice situation awareness in naturalistic decision making. *International Journal of Human Computer Studies, 45*, 579–597.

Rauch, S. L., Shin, L. M., Segal, E., Pitman, R. K., Carson, M. A., Whalen, P. J., et al. (2003). Selectively reduced regional cortical volumes in posttraumatic stress disorder. *Neuroreport, 14*(7), 913–916.

Rempel, J., Holmes, J., & Zanna, M. (1985). Trust in close relationships. *Journal of Personality & Social Psychology, 49*(1), 95–112.

Rencoret, N., Stoddard, A., Haver, K., Taylor, G., & Harvey, P. (2010, July). *Haiti earthquake response: Context analysis.* Active Learning Network for Accountability and Performance in Humanitarian Action. Retrieved January 3, 2012, from https://reliefweb.int/report/haiti/haiti-earthquake-response-context-analysis-july-2010

Renshaw, K. D. (2011). An integrated model of risk and protective factors for postdeployment PTSD symptoms in OEF/OIF era combat veterans. *Journal of Affective Disorders, 128*(3), 321–326.

Report by the Select Bipartisan Committee to Investigate the Preparation for and Response to Hurricane Katrina (2006). A failure of Initiative: Final Report of the Select Bipartisan Committee to Investigate the Preparation for and Response to Hurricane Katrina. Available via the World Wide Web: http://www.gpoacess.gov/congress/index.html

Rich, B. L., LePine, J. A., & Crawford, E. R. (2010). Job engagement: Antecedents and effects on job performance. *Academy of Management Journal, 53*(3), 617–635.

Roach, M. (2016). *Grunt: The curious science of soldiers at war.* New York, NY: Norton.

Rose, R. M., Bourne, P. G., Poe, R. O., Mougey, E. H., Collins, D. R., & Mason, J. W. (1969). Androgen responses to stress: II. Excretion of testosterone, epitestosterone, androsterone and etiocholanolone during basic combat training and under threat of attack. *Psychosomatic Medicine, 131*, 418–436.

Roseman, I., Wiest, C., & Swartz, T. (1994). Phenomenology, behaviors and goals differentiate emotions. *Journal of Personality and Social Psychology, 67*(2), 206–221.

Rosen, S. P. (1991). *Winning the next war: Innovation and the modern military.* Ithaca, NY: Cornell University Press.

Ross, K. G., Klein, G. A., Thunholm, P., Schmitt, J. F., & Baxter, H. C. (2004, July–August). The recognition-primed decision model. *Military Review*, 6–10.

Ruark, G. A., Orvis, K. L., Horn, Z., & Langkamer, K. L. (2009). Trust as defined by U.S. Army soldiers. *Human Factors and Ergonomics Society Annual Meeting Proceedings, 53*(26), 1924–1928.

Rubin, R. T., Miller, R. G., Arthur, R. J., & Clark, B. R. (1970). Differential adrenocortical stress responses in naval aviators during aircraft carrier landing practice. *Psychological Reports, 26*, 71–74.

Rubin, M., Shvil, E., Papini, S., Chhetry, B. T., Helpman, L., Markowitz, J. C., . . . Neria, Y. (2016). Greater hippocampal volume is associated with PTSD treatment response. *Psychiatry Research: Neuroimaging, 252*, 36–39. https://doi.org/10.1016/j.pscychresns.2016.05.001

Saavedra, R., Earley, P. C., & Van Dyne, L. (1993). Complex interdependence in task-performing groups. *Journal of Applied Psychology, 78*(1), 61–72.

Salas, E., Rosen, M. A., Burke, S., Goodwin, G. F., & Fiore, S. M. (2006). The making of a dream team: When expert teams do best. In K. A. Ericsson, N. Charness, P. J. Feltovich, & R. R. Hoffman (Eds.), *The Cambridge handbook of expertise and expert performance: Its development, organization, and content* (pp. 439-453). Cambridge, UK: Cambridge University Press.

Salas, E., Sims, D. E., & Burke, C. S. (2005). Is there a "big five" in teamwork? *Small Group Research, 36*(5), 555-599.

Scales, R. (2009, October). Return of the Jedi. *Armed Forces Journal,* 22.

Schmitt, J. F. (1994). Mastering tactics. Quantico, VA: Marine Corps Association.

Schmitt, J. F., & Klein, G. (1996). Fighting in the fog: Dealing with battlefield uncertainty. *Marine Corps Gazette, 80,* 62-69.

Schmitt, J. F., & Klein, G. (1999a). How we plan. *Marine Corps Gazette, 83*(10), 18-26.

Schmitt, J. F., & Klein, G. (1999b). A recognition planning model. *Proceedings of the 1999 Command and Control Research and Technology Symposium, 1,* 510-521.

Scholten, M. (2002). Conflict-mediated choice. *Organizational Behavior and Human Decision Processes, 88,* 683-718.

Scholten, M., & Sherman, S. (2006). Tradeoffs and theory: The double-mediation model. *Journal of Experimental Psychology: General, 135,* 237-261.

Schön, D. A. (1983). *The reflective practitioner.* New York, NY: Basic Books.

Schwartz, S. H. (1992). Universals in the content and structure of values: Theoretical advances and empirical tests in 20 countries. *Advances in Experimental Social Psychology, 25,* 1-65.

Schwartz, S. H. (2005). Basic human values: Their content and structure across cultures. In A. Tamayo & J. Porto (Eds.), *Valores e trabalho* [Values and work] (pp. 21-55). Brasilia: Editoa Vozes.

Schwartz, S. H. (2009). Basic values: How they motivate and inhibit prosocial behavior, In M. Mikulincer & P. Shaver (Eds.), *Herzliya Symposium on Personality and Social Psychology, Vol. 1.* Washington, DC: American Psychological Association Press.

Scranton, R., & Gallagher, M. (Eds.). (2013). *Fire and forget: Short stories from the long war.* Boston, MA: Da Capo.

Selye, H. (1936). Stress: A syndrome produced by diverse nocuous agents. *Nature, 138,* 32-32.

Shafir, E., Simonson, I., & Tversky, A. (1993). Reason-based choice. *Cognition, 49,* 11-36.

Shamir, B., Zakay, E., Breinin, E., & Popper, M. (1998). Correlates of charismatic leader behavior in military units: Subordinates' attitudes, unit characteristics, and superiors' appraisals of leader performance. *Academy of Management Journal, 41*(4), 384-409.

Shanteau, J. (1997). Competence in experts: The role of task characteristics. *Organizational Behavior and Human Decision Processes, 53,* 252-266.

Shattuck, L. G., Miller, N. L., & Kemmerer, K. E. (2009). Tactical decision making under conditions of uncertainty: An empirical study. *Human Factors and Ergonomics Society Annual Meeting Proceedings, 53*(4), 242-246.

Shaw, D., & Thomson, J. (2013). Spirituality and ethical consumption. *European Journal of Marketing, 47*(4), 557-573.

Shay, J. (1995). *Achilles in Vietnam: Combat Trauma and the Undoing of Character.* New York, NY: Scribner.

Shepard, R. W., & Fare, R. (1974). The law of diminishing returns. *Zeitschrift fur Nationalokonomie, 34,* 69-90.

Shoffner, W. A. (2000). *The military decision-making process: Time for a change*. Fort Leavenworth, KS: School of Advanced Military Studies, United States Army Command and General Staff College.

Shortland, N. D., & Alison, L. J. (2015). War stories: A narrative approach to understanding military decisions. *The Military Psychologist, 30*(2).

Shortland, N. D., & Bohannon, J. (2014). Civilian casualties in Afghanistan. *Science, 345*, 731-733.

Shortland, N. D., & Hilland, C. (2013). *Getting left of green-on-blue: Insider attacks in Afghanistan: Final report prepared for the United Kingdom Ministry of Defense*. Boston, MA: Center for Terrorism and Security Studies.

Siebold, G. L. (2007). The essence of military group cohesion. *Armed Forces and Society, 33*(2), 286-295.

Simmons, J. M. (1997). *Marketing higher education: Applying a consumption value model to college choice*. PhD dissertation, Marquette University, Australia.

Simon, H. A. (1955). A behavioral model of rational choice. *Quarterly Journal of Economics, 69*, 99-118.

Simon, H. A. (1956). Rational choice and the structure of the environment. *Psychological Review, 63*, 129-138.

Simon, H. A. (1958). Review: "The Decision-Making Schema": A Reply. *Public Administration Review, 18*, 60-63.

Simon, H. (1976). *Administrative Behavior; A study of Decision-Making Processes in Administrative Organizations, 3rd*. New York: Free Press.

Simons, T., & Peterson, R. (2000). Task conflict and relationship conflict in top management teams: The pivotal role of intragroup trust. *Journal of Applied Psychology, 85*(1), 102-111.

Simonson, I., & Tversky, A. (1992). Choice in context: Tradeoff contrast and extremeness aversion. *Journal of Marketing Research, 29*, 281-295.

Smith, S., & Fordy, G. (in press). The effect of load carriage on cognitive performance: Implications for military task performance during dismounted operations. *Military Psychology*.

Sniezek, J. A., & Van Swol, L. M. (2001). Trust and expertise in a judge advisor system. *Organizational Behaviour and Human Decision Processes, 82*, 288-307.

Spranca, M., Minsk, E., & Baron, J. (1991). Omission and commission in judgment and choice. *Journal of Experimental Social Psychology, 27*, 76-105.

Stanton, N. A. (2011). *Trust in Military Teams*. London, UK: Routledge.

Starcke, K., & Brand, M. (2016). Effects of stress on decisions under uncertainty: a meta-analysis. *Psychology Bulletin, 142*, 909-933. doi:10.1037/bul0000060

Staw, B. M. (1976). Knee-deep in the big muddy: A study of escalating commitment to a chosen course of action. *Organizational Behavior and Human Performance, 16*(1), 27-44.

Staw, B. M., & Ross, J. (1989). Understanding behavior in escalation situations. *Science, 246*(216), 220.

Svenson, O. (1979). Process descriptions of decision making. *Organizational Behaviour and Human Performance, 26*, 86-112.

Sweller, J. (2004). Instructional design consequences of an analogy between evolution by natural selection and human cognitive architecture. *Instructional Science, 32*, 9-31.

Sweller, J., van Merriënboer, J. J. G., & Paas, F. (1998). Cognitive architecture and instructional design. *Educational Psychology Review, 10*, 251-296.

Taleb, N. N. (2008). *The black swan: The impact of the highly improbable.* London, UK: Penguin.

Tanielian, T., & Jaycox, L. (Eds.). (2008). *Invisible wounds of war: Psychological and cognitive injuries, their consequences, and services to assist recovery.* Santa Monica, CA: RAND Corporation.

Taylor, M. K., Reis, J. P., Sausen, K. P., Padilla, G. A., Markham, A. E., Potterat, E. G., & Drummond, S. P. (2008). Trait anxiety and salivary cortisol during free living and military stress. *Aviation Space and Environmental Medicine, 79*(2), 129–135.

Taylor, M. K., Sausen, K. P., Mujica-Parodi, L. R., Potterat, E. G., Yanagi, M. A., & Kim, H. (2007). Neurophysiologic methods to measure stress during survival, evasion, resistance, and escape training. *Aviation Space and Environmental Medicine, 78*(5), 224–230.

Tetlock, P. E. (2003). Thinking about the unthinkable: Coping with secular encroachments on sacred values. *Trends in Cognitive Science, 7*(7), 320–324.

Tetlock, P. E., & Boettger, R. (1989). Accountability: A social magnifier of the dilution effect. *Journal of Personality and Social Psychology, 57,* 388–398.

Tetlock, P. E., Kristel, O. V., Elson, S. B. Lerner, J. S., & Green, M. C. (2000). The psychology of the unthinkable: Taboo trade-offs, forbidden base rates, and judgment and heretical counterfactuals. *Journal of Personality and Social Psychology, 78*(5), 853–870.

The Research and Technology Organisation (RTO) of NATO. (2005). *Military Command Team Effectiveness: Model and Instrument for Assessment and Improvement.* RTO/NATO: Neuilly-sur-Seine Cedex, France.

Thunholm, P. (2005). Planning under time pressure: An attempt toward a prescriptive model of military tactical decision making. In H. Montgomery, R. Lipshitz, & B. Brehmer (Eds.), *How professionals make decisions* (pp. 43–56). London, UK: Erlbaum.

Tibbett, T. P., & Ferrari, J. R. (2015). The portrait of the procrastinator: Risk factors and results of an indecisive personality. *Personality and Individual Differences, 82,* 175–184.

Treviño, L. K. (1986). Ethical decision making in organizations: A person–situation interactionist model. *Academy of Management Review, 11*(3), 601–617.

Tversky, A. (1975). A Critique of Expected Utility Theory: Descriptive and Normative Considerations. *Erkenntnis, 9*(2), 163–173.

Tversky, A., & Kahneman, D. (1973). Availability: A heuristic for judging frequency and probability. *Cognitive Psychology, 5,* 207–232.

Tversky, A., & Shafir, E. (1992). Choice under conflict: The dynamics of deferred decision. *Psychological Science, 6,* 358–361.

Tversky, A., & Simonson, I. (1993). Context-dependent preferences. *Management Science, 39,* 1179–1189.

United States Army. (1910). *Field Service Regulations.* Washington, D.C.: War Department.

United States Army. (1997). *Staff organization and operations* (Field Manual 101). Washington, DC; Headquarters, Department of the Army.

United States Army. (2003). *Mission command: Command and control of Army forces* (Field Manual 6-0). Washington, DC: Headquarters, Department of the Army.

United States Army. (2009). *Combat and operational stress control manual for leaders and soldiers* (Field Manual 6-22.5). Washington, DC: Headquarters, Department of the Army.

United States Army. (2012). *Army leadership* (ADRP 6-22). Washington, DC: Headquarters, Department of the Army.

United States Army. (2015). *US Army Human Dimension Strategy*. Washington DC: United States Army Headquarters.

United States Army/Marine Corps. (2014). *Counterinsurgency field manual* (US Army Field Manual No. 3-24/Marine Corps Warfighting Publication No. 3-33.5). Washington, DC: Headquarters, Department of the Army.

US Army Combined Arms Center. (2015). The United States Army: Washington, D.C. http://data.cape.army.mil/web/character-development-project/repository/human-dimension-strategy-2015.pdf

Vaernes, R., Ursin, H., Darragh, A., & Lambe, R. (1982). Endocrine response patterns and psychological correlates. *Journal of Psychosomatic Research*, 26, 123–131.

van den Heuvel, C., Alison, L., & Crego, J. (2012). How uncertainty and accountability can derail strategic "save life" decisions in counter-terrorism simulations: A descriptive model of choice deferral and omission bias. *Journal of Behavioral Decision Making*, 25, 165–187.

van den Heuvel, C., Alison, L. J., & Power, N. (2014). Coping with uncertainty: Police strategies for resilient decision-making and action implementation. *Cognition, Technology & Work*, 16(1), 25–45.

Wall, V., & Nolan, L. N. (1986). Perceptions of inequity, satisfaction, and conflict in task-oriented groups. *Human Relations*, 39(11), 1033–1052.

Wansink, B., Payne, C. R., & North, J. (2007). Fine as North Dakota wine: Sensory expectations and the intake of companion foods. *Physiology and Behavior*, 90(5), 712–716.

Waring, S., Alison, L., Cunningham, S., & Whitfield, K. C. (2013). The impact of accountability on motivational goals and the quality of advice provided in crisis negotiations. *Psychology, Public Policy and Law*, 19(2), 137–150.

Wason, P. (1968). Reasoning about a rule. *Quarterly Journal of Experimental Psychology*, 20, 273–281.

Weems, C. F., Pina, A. A., Costa, N. M., Watts, S. E., Taylor, L. K., & Cannon, M. F. (2007). Pre-disaster trait anxiety and negative affect predict posttraumatic stress in youths after Hurricane Katrina. *Journal of Consulting and Clinical Psychology*, 75, 154–159.

Weisaeth, L., Knudsen, O., Jr., & Tonnessen, A. (2002). Technological disasters: Crisis management and leadership stress. *Journal of Hazardous Materials*, 93(1), 33–45.

White, S. S., Mueller-Hanson, R. A., Dorsey, D. W., Pulakos, E. D., Wisecarver, M. M., Deagle, E. A., & Mendini, K. G. (2005). *Developing adaptive proficiency in Special Forces officers* (ARI Research Report No. 1831). Arlington, VA: US Army Research Institute for the Behavioral and Social Sciences. (DTIC No. ADA432443)

Wilson, D. C., Butler, R. J., Cray, D., Hickson, D. J., & Mallory, G. R. (1986). Breaking the bounds of organization in strategic decision making. *Human Relations*, 39(4), 309–332.

Wilson, T. D., Wheatley, T. P., Meyers, J. M., Gilbert, D. T., & Axsom, D. (2000). Focalism: A source of durability bias in affective forecasting. *Journal of Personality and Social Psychology*, 78, 821–836.

Witte, E. (1972). Field research on complex decision-making processes—The phase theorem. *International Studies of Management and Organization*, 2(2), 156–182.

Wolfe, J., & Kimerling, R. (1997). Gender issues in the assessment of posttraumatic stress disorder. In J. P. Wilson & T. M. Keane (Eds.), *Assessing psychological trauma and PTSD* (pp. 192–238). New York: Guilford Press.

Woodzicka, J. A., & LaFrance, M. (2001). Real versus imagined gender harassment. *Journal of Social Issues*, 57, 15–30.

Wong, K. F. E., & Kwong, J. Y. Y. (2007). The role of anticipated regret in escalation of commitment. *Journal of Applied Psychology, 92*(2), 545–554.

Wong, L., Kolditz, T. A., Millen, R. A., & Potter, T. M. (2003). *Why they fight: Combat motivation in the Iraq War.* Carlisle, PA: Strategic Studies Institute.

Workman, N. M., Lesser, M. F., & Kim, J. (2007). An exploratory study of cognitive load in diagnosing patient conditions. *International Journal of Qualitative Health Care, 19*(3), 127–133.

Yates, J. F., & Estin, P. A. (1998). Decision making, In W. Bechtel & G. Graham (Eds.), *A companion to cognitive science* (pp. 186–196). Malden, MA: Blackwell.

Yates, J. F., & Patalano, A. L. (1999). Decision making and aging. In D. C. Park, R. W. Morrell, & K. Shifren (Eds.), *Processing of medical information in aging patients: Cognitive and human factors perspectives* (pp. 31–54). Mahwah, NJ: Erlbaum.

Yates, J. F., & Stone, E. R. (1992). The risk construct. In J. F. Yates (Ed.), *Risk-taking behavior* (pp. 1–26). New York, NY: Wiley.

Yates, J. F., Veinott, E. S., & Patalano, A. L. (2003). Hard decisions, bad decisions: On decision quality and decision aiding. In S. L. Schneider & J. C. Shanteau (Eds.), *Emerging perspectives on judgment and decision research* (pp. 13–63). New York, NY: Cambridge University Press.

Yetiv, S. (2004). *Explaining foreign policy: U.S. decision-making and the Persian Gulf War.* Baltimore, MD: Johns Hopkins University Press.

Zakay, D. (1985). Post-decisional confidence and conflict experienced in a decision process. *Acta Psychologica, 58*, 75–80.

Zakay, D., & Tsal, Y. (1993). The impact of using forced decision making strategies on post-decisional confidence. *Journal of Behavioral Decision Making, 6*, 53–68.

Zeelenberg, M. (1999). Anticipated regret, expected feedback and behavioral decision-making. *Journal of Behavioral Decision Making, 12*, 93–106.

Zeelenberg, M., & Beattie, J. (1997). Consequences of regret aversion: 2. Additional evidence for effects of feedback on decision making. *Organizational Behavior and Human Decision Processes, 72*, 63–78.

Zeelenberg, M., van den Bos, K., van Dijk, E., & Pieters, R. (2002). The inaction effect in the psychology of regret. *Journal of Personality and Social Psychology, 82*(3), 314–327.

Zeleny, M. (1976). The theory of the displaced ideal. In M. Zeleny (Ed.), *Multiple criteria decision making* (pp. 155–206). New York, NY: Springer-Verlag.

Zimbardo, P. G. (1969). The human choice: Individuation, reason, and order versus deindividuation, impulse, and chaos. In W. D. Arnold & D. Levine (Eds.), *Nebraska Symposium on Motivation* (pp. 237–307). Lincoln, NE: University of Nebraska.

Zsambok, C. E. (1997). Naturalistic decision making: Where are we now? In C. E. Zsambok & G. A. Klein (Eds.), *Naturalistic decision making* (pp. 3–16). Mahwah, NJ: Erlbaum.

INDEX

accountability, 135, 178
  defined, 45
  plan formulation and, 78–79
  SAFE-T model on, 44–47
action, 37, 101–2
advisors (hardness factor 7), 3, 4, 7
Afghanistan, ix, 11, 189
  accountability in, 46
  decision-making in based on previous conflict, xiii
  in extremis decision-making in, 140–46
  interoperability in, 122–28, 131–35
  length of war in, xix
  military decision-making in, 15–17
  Operation Enduring Freedom, 47–49, 164
  plan execution in, 102–5
  plan formulation in, 79–83, 94–99
  PTSD in soldiers, 170–71
  published literature on, xix
  recognition planning model and, 27–28, 31–32
  situation awareness in, 56–60, 62–64, 65–73
  team learning in, 120
  uncertainty in, 48–49
Afghanistan National Security Forces (ANSF), 97–98
Afghan National Army (ANA), 48–49, 64, 118, 140, 141, 144, 148–49
Afghan National Police (ANP), 122–28
agreeableness, 180–81, 182–83
Air Force Research Laboratory's 711th Human Performance Wing, 185–86
airplane crash simulation, 135–36
air traffic controllers, 6
American Psychiatric Association, 164
amygdala, 166
ANA. *See* Afghan National Army

anonymization, 45
ANP. *See* Afghan National Police
anticipatory regret, 177–78
  accountability and, 44, 45–46
  defined, 43
  plan formulation and, 78–79
  SAFE-T model on, 43–44
anxiety, 154
apparent actions, 42
approach goals, 88–90
argumentation, 87–88
Aristotle, 99
Army Human Dimension Strategy, 47–48, 159
Army Leadership Doctrine, 113–14
Army Research Institute, 24, 151–52, 189
Army Research Lab's Advanced Training and Simulation Division, 187
attribute importance, 88, 89–90, 92, 98
avoidance goals, 88–90

"backwards law," 179
Barker, Col. (interview subject), 65–69, 162, 163, 172–73, 181
berserking, 110n1
biases
  cognitive, 74–75
  confirmation, 68, 78–79
  durability, 43–44
  hindsight, 13
  impact, 43–44
  in military decision-making process, 21–22
  omission, 45–46
big five personality factors, 180–81
binary decisions, 86
"black swans," 33, 177
blue light services, 88–89
bomb paradigm, 5–6

Boston Marathon bombers, 184–85
Boxing Day tsunami, 7, 41
Buncefield fire, 7
Buridan, Jean, 99, 100
Bush, George H. W., 106

Cabinet Office Briefing Room A (COBRA), ix
category-based trust, 117
CDM. *See* critical decision method interviews
challenge (stressor as), 155
Chase, Lt. Col. (interview subject), 128–30
chess analogy, 24–25, 28
Chilcot investigation, x
choice
   conflict and, 85–92, 99–100
   course of action compared with, xv*n*2
   plan formulation and, 77–78, 83–92, 99–100
   soldiers' ability to make, 84–85, 177
choice deferral, 38
"choking," 45
civilians
   difficulty of leading in combat zones, ix
   minimizing casualties in, 97–98
   soldiers compared with on personality, 180–83
   soldiers' culture compared to culture of, 183–84
clarity (hardness factor 5), 3, 4
clearing operations, 141
Clinton, Hillary, 10–11
cognitive appraisal, 155
cognitive biases, 74–75
cognitive capacity, 147
cognitive conflict, 78–79, 82, 83, 96–97, 99–100
cognitive control, 149–50, 152, 183
cognitive load, 41, 114
   development of theory, 147
   extraneous, 148–49
   in extremis decision-making and, 147–52, 158
   germane, 148–49
   intrinsic, 147–49

Collins, Timothy, ix–xiv
combat exhaustion, 164
commitment, 77–79
confidence, 7, 8–9, 59, 74–75, 82, 86
confirmation bias, 68, 78–79
conflict
   cognitive, 78–79, 82, 83, 96–97, 99–100
   decisional (*see* decisional conflict)
   double-mediation model of, 88–89
   emotional (*see* emotional conflict)
   process, 106–7, 112
   relationship (*see* relationship conflict)
   task (*see* task conflict)
   of values, 105–7
conscientiousness, 180–81
coping, 155
"corporal's war," ix
corrective component of team learning, 120–21
correct rejection, 162–63
cortisol, 154–55
"courageous restraint," xi
course of action (CoA), xv*n*2, 18–19, 35
cowardice, xi
critical decision method (CDM) interviews, xvii–xix
cultural factors, 11, 46–47
   armed forces *vs.* civilian, 183–84
   situation awareness and, 53, 64, 74–75

damned if you do; damned if you don't decisions, 11
Davids, Lt. Col. (interview subject), 12, 131–35, 185–86
decisional conflict
   choice and, 85–92, 99–100
   defined, 37
   primary factors in, 1–14
decision avoidance, 38–39, 43
   defined, 34, 38
   forms taken by, 38
decision error
   defined, 12–13
   outcomes *vs.* processes in, 12–13
   plan-continuation, 64
   SAFE-T model and, 49–50
   situation awareness and, 64–73, 76

Index

decision inertia, xv–xvi, xvii–xviii, 37–40, 41, 188–89
  decision avoidance contrasted with, 38–39
  defined, 34
  diminishing returns concept and, 54
  indecision contrasted with, 37–38
  in extremis decision-making and, 143, 151–52, 156, 158
  interoperability and, 135–36
  lack of research on, xvii–xviii
  least-worst decisions and, 178–79
  plan formulation and, 88–90, 93, 98, 99–100
  prevalence of, 41
  resistance to, xviii
  SAFE-T model and, 37, 42, 51
  signs of, xvii–xviii
  the tunnel scenario, 39–40
  values and, 93, 98
decision-making
  causes of damage to, xi–xiii
  if A then B, 83
  in extremis (*see* in extremis decision-making)
  interviews and interview method on, xvii–xix
  lab-based research on, 12–13
  military (*see* military decision-making)
  rational-cognitive approaches to, 17, 37–38
  rational comprehensive model of, 21
  recognition-primed model of (*see* recognition-primed model of decision-making)
  stress effects on, 172–73
  traditional, 20–21
  trimodal theory of, 77–78
decisions
  binary, 86
  damned if you do; damned if you don't, 11
  defined, 2, 37
  hard (*see* hard decisions)
  intervene/don't intervene, 11
  least-worst (*see* least-worst decisions)
  multiattribute, 12–13, 20, 176–77
  shoot/don't shoot, 10, 11
  trade-offs in (*see* trade-offs)
Defense Science and Technology Laboratory at Porton Down, 150
Dempsey, Martin, 113–14
deontological behavior, 182–83
desirability, 76, 83
diagnosis of problem. *See* inability to diagnose problem
Diagnostic and Statistical Manual of Mental Disorders, 164
diminishing returns concept, 54
disengaging, 41
disjuncted commitment, 77–78
doctrine, 23, 135, 188–89
  plan execution and, 104
  plan formulation and, 78–80, 82, 83, 99–100
donkey's dilemma, 99
double-mediation model of conflict, 88–89
drones/drone pilots, 61–64, 69–73
durability bias, 43–44

economic rationality, 21
effective cognitive load. *See* germane cognitive load
embedded journalism, 46
emergency services, 183–84, 188
emotion, 7, 8–9, 82
emotional conflict. *See also* relationship conflict
  plan execution and, 104, 106–7
  plan formulation and, 78–79
emotional stability, 180–81, 182–83
endogenous uncertainty, 47, 48–49, 121–22, 177–78
error. *See* decision error
Evans, AFC R. (interview subject), 62–64
exogenous uncertainty, 47, 48–49, 177–78
  defined, 121–22
  interoperability and, 121–22, 127
expertise and experience
  naturalistic decision-making and, 24
  plan formulation and, 84, 99–100
  situation awareness and, 60–61
exposure therapy, 167

extraneous cognitive load, 148–49
extraversion, 180–81

failures to act, xv–xvi, 37–39, 41, 135–36
false alarms, 162–63
fear-extinction memories, 167
Fick, Nathanial (interview subject), 150–51
fight-or-flight response, 138–39
firefighters, 26–27, 37, 183–84, 188
flashbulb memories, 165, 166
fMRI. *See* functional magnetic resonance imaging
force protection, 89–92, 96–97, 99–100
Fort Leavenworth Battle Command Battle Laboratory (BCBL), 30
frontal lobes, 151, 152
full commitment, 77–78
functional magnetic resonance imaging (fMRI), 151, 166

game theory, 157
gender and PTSD, 168
genetic factors and PTSD, 165, 173
germane cognitive load, 148–49
glucocorticoid hormones, 154–55
Green, Lt. Col. S. (interview subject), 79–83, 102–6, 112, 114, 117, 148–49
green army, x
groupthink, 106
Gulf War, 46, 153

Haditha triad, 128–29
Haqqani network, 65
hard decisions, xviii, 1–8
    characteristics of, 2–3
    environmental factors in, 5–8
    office party scenario, 3–4
    supercategories of, 2–3
hardness factor 1 (outcomes--serious), 2, 4
hardness factor 2 (options), 2, 4
hardness factor 3 (process--onerous), 2, 4, 7
hardness factor 4 (possibilities), 2, 4, 7
hardness factor 5 (clarity), 3, 4
hardness factor 6 (value), 3, 4
hardness factor 7 (advisors), 3, 4, 7
harm-loss, 155
heuristics, 21–22, 23, 74–75
hindsight bias, 13

hippocampus, 166, 167
hits, 162–63
HM (epilepsy patient), 166
hostage negotiation simulations
    accountability in, 45–46
    endogenous uncertainty in, 48–49, 121–22
    plan formulation in, 87, 88
human in the loop, 159
hurdles of interoperability, 127
Hurricane Katrina, 41, 154

IEDs. *See* improvised explosive devices
if A then B decision-making, 83
ignoring (of a problem), 41
immersive training, 154
impact bias, 43–44
implementation failure, 34, 39, 41
improvised explosive devices (IEDs), xvi, 31–32, 123, 127, 131, 161, 163
    in extremis decision-making and, 140, 141, 142–44, 145–46, 156–57
    situation awareness and, 61–64
inability to diagnose problem, 7, 8–9, 82
indecision, 39
    decision inertia contrasted with, 37–38
    forms taken by, 37–38
    SAFE-T model and, 34
    values and, 98
inertia. *See* decision inertia
in extremis decision-making, 103, 137, 138–59, 189
    cognitive load and, 147–52, 158
    defined, 138
    difficulty of studying, 138
    PTSD and, 154, 165
    recognition-primed model and, 156–58
    stress in, 153–56
information volume, 7, 8–9, 81–82
International Security Assistance Force (ISAF), 48–49, 97–98
    in extremis decision-making and, 140
    interoperability and, 122–28
interoperability, 120–36
    decision inertia and, 135–36
    examples of, 122–30, 131–35
    exogenous uncertainty and, 121–22, 127
    hurdles of, 127
    as team learning, 120–21

intervene/don't intervene decisions, 11
intrinsic cognitive load, 147–49
intuition, 24, 177
Iran, xiv
Iran-Contra scandal, 106
Iraq, ix, 11, 189
  accountability in, 46
  Chilcot investigation, x
  comparison of US and UK military in, x
  decision-making in based on previous conflict, xiii
  Gulf War, 46, 153
  interoperability in, 128–30
  Operation Iraqi Freedom, 47–49, 115, 164
  plan formulation in, 91
  PTSD in soldiers, 170–71
  published literature on, xix
  relationship conflict in, 107–11
  team learning in, 120
  uncertainty in, 48–49
  weapons of mass destruction claim, 106
Iraqi Defense Forces, 23, 48–49
ISAF. *See* International Security Assistance Force
Israeli Defense Forces, 114, 116

James, Maj. (interview subject), 123–28, 130
Joint Vision 2020, 26–27

Karzai, Hamid, 79, 103–4
Kuwait, 106

least-worst decisions, xv, 9–11, 175, 186–87, 188–89
  "black swans" comparison, 33
  decision inertia and, 178–79
  defining characteristic of, 9
  in extremis (*see* in extremis decision-making)
  interoperability as, 126, 127
  lack of research on, xvi
  personality and, 182–83
  plan execution and, 118
  plan formulation and, 76–77, 80, 82, 93, 95, 96, 97–98, 99
  PTSD and (*see* post-traumatic stress disorder)
  RPD/RPM and, 31–33
  science of selecting, 34–52 (*see also* SAFE-T model)
  situation awareness and, 53–54
  the "subtle art of not giving a f*ck" and, 179–80
  trust and, 116
  values and, 93, 94, 95, 96, 97–98, 105, 119, 169–70, 177, 178–79, 189
  vividness of recall, xviii–xix
Level 1 situation awareness, 55–56, 59, 60, 61–62, 63, 64–65, 73–75
Level 2 situation awareness, 55–56, 59, 60, 61, 63, 73–75
Level 3 situation awareness, 55–56, 59, 60, 61, 63, 74–75
lexis of critical incidents (LOCI), 186–87
London bombings. *See* 7/7 London bombings
London Olympic Games of 2012, 7
Los Angeles Times, 69
lost opportunity hypothesis, 43

Marshall, Col. S. L. A. (interview subject), 53
matching, 77–78
  defined, 77–78
  example of, 79–83
materiel solutions, 185–86, 187
McChrystal, Stanley, 97–98
MDMP. *See* military decision-making process
medial prefrontal cortex, 166–67
memory
  fear-extinction, 167
  flashbulb, 165, 166
  working, 45, 74–75, 147–48
meta-analysis, 165, 168, 172–73
MIEs. *See* morally injurious experiences
military decision-making, 15–33. *See also* military decision-making process; naturalistic decision-making; recognition planning model of military decision-making
  process of, 175–78
  types of approaches to, 15–17

military decision-making process
(MDMP), 17–23, 50–52
   heuristics and biases in, 21–22
   practical and theoretical issues, 20
   purpose of, 17–18
   recognition planning model and, 26–27, 28, 29, 30
   SAFE-T model and, 34, 35, 36, 37, 49–52
   traditional decision-making comparison, 20–21
   uncertainty and, 22–23
Minehounds, 141–43, 146, 148–49, 155
MiniDome environment, 187
misattribution and misinterpretation hypothesis, 112–13
misses, 162–63
Mission Command Center of Excellence at NSRDEC, 187
monitoring, 120–21, 185–86
moral cowardice, xi
Moral Injury Questionnaire--Military Version (MIQ-M), 169–70
morally injurious experiences (MIEs), 168–70, 171
moral psychology, 182–83
multiagency communication, 7, 8–9, 82
multiattribute decisions, 12–13, 20, 176–77

Natick Soldier Research, Development and Engineering Center (NSRDEC), 148, 185–86, 187
National Vietnam Veterans Readjustment Study (NVVRS), 164
naturalistic decision-making (NDM), 17, 23–24, 50, 177–78
   characteristics focused on by, 24
   decision inertia not addressed by, 37
   development of, 24
   errors and, 13
   situation awareness and, 60
naturalistic research, 35, 45–46, 48–49, 88–89, 135–36
NDM. *See* naturalistic decision-making
neuroticism, 180–81, 182
New York Times, 189
non-actions, 42
non-materiel solutions, 185–86, 187

North Atlantic Treaty Organization, 48–49
Northern Ireland, ix
Norwegian Military Academy, 151
NSRDEC. *See* Natick Soldier Research, Development and Engineering Center

Obama, Barack, 10–11
office party scenario, 3–4
Olympic Games of 2012, 7
omission, 38
omission bias, 45–46
OODA loop, 49–50
openness to experience, 180–81, 182–83
Operation Desert Storm, 46
Operation Enduring Freedom, 47–49, 164
Operation Iraqi Freedom, 47–49, 115, 164
Operation Theseus, 7–8
optimization, 9–10, 185–86, 187
options (hardness factor 2), 2, 4
outcome mutability, 43
outcomes, xv–, 5, 8–9
   decisions not always correlated to, 1, 68
   hard decisions and, 1, 4–5
   possible (hardness factor 4), 2, 4, 7
   processes *vs.*, 12–13
   serious (hardness factor 1), 2, 4

paramedics, 183–84, 188
Park, Corporal (interview subject), 107–11
Patton, George S., 27
peritraumatic dissociation, 165
personality of soldiers *vs.* civilians, 180–83
person-based trust, 115
persons indicted for war crimes in the former Yugoslavia (PIFWCs), ix
Petraeus, General, 29, 125
physiological load, 149–50, 151–52
plan-continuation errors, 64
plan execution (PE), 35, 101–19
   accountability and, 45–46
   defined, 35
   OODA loop and, 49–50
   trust and, 101, 112–18
   uncertainty and, 47–48
   values and, 105–7, 119

plan formulation (PF), 35, 76–100
  accountability and, 45–46
  choice in (*see under* choice)
  defined, 35
  OODA loop and, 49–50
  sacred values and (*see* sacred values)
  uncertainty and, 47–48, 82
police, 183–84, 188
policy, 78–79, 83, 99–100, 104
positive identification (PID)
  plan execution and, 102, 104, 114
  plan formulation and, 81, 82–83
  situation awareness and, 69–70, 71, 72, 73–74
post-battle experiences, 171
post-traumatic stress disorder (PTSD), xviii–xix, 160–74
  exposure therapy for, 167
  in extremis decision-making and, 154, 165
  history in the military context, 164
  morally injurious experiences and, 168–70, 171
  neuroscience of, 166–67
  origin of diagnosis, 164
  predictive factors for development of, 165, 168
  prevalence in post-combat soldiers, 164
  resilience to, 170–73
  symptoms of, 164–65
prediction, 185–86, 187
preference uncertainty, 85–86
prefrontal cortex, 150, 166–67
primary cognitive appraisal, 155
process conflict, 106–7, 112
process-driven model of stress response, 155
processes
  of military decision-making (*see* military decision-making process)
  onerous (hardness factor 3), 2, 4, 7
  outcomes *vs.*, 12–13
processive stressors, 172
Project Afghan (website), 180*n*2
PTSD. *See* post-traumatic stress disorder

al-Qa'ida, 91, 128–29, 130

Rangers, 140, 156
rational-cognitive approaches to decision-making, 17, 37–38
rational comprehensive model of decision-making, 21
RAWFS heuristic, 23
reassessment, 83
recognition planning model of military decision-making (RPM), 25–33
  advantages of, 29
  important flaw in, 30
  least-worst decisions and, 31–33
  MDMP and, 26–27, 28, 29, 30
recognition-primed model of decision-making (RPD), 10, 24–25, 31, 50–52, 101–2, 176–78, 186, 188–89
  choice not a factor in, 77, 85
  decision inertia not addressed by, 37
  in extremis, 156–58
  least-worst decisions and, 31–33
  plan formulation and, 77, 83, 84, 85
  reassessment and, 83
  SAFE-T model compared with, 37
  situation awareness and, 61
regret. *See* anticipatory regret
regulators, xii–xiii
relationship conflict, 106–11, 117–18. *See also* emotional conflict
  causes of, 106–7
  in Iraq, 107–11
  task conflict and, 112, 119
  trust and reduction of, 117
Republican Movement, xiii
research
  future questions for emergency services, 188
  future questions for military, 185–87
Revolutionary War, 17–18
Rockwell, Floyd, 141–42
role-based trust, 117
role confusion, 122, 177–78
role understanding, 48–49
routine trade-off, 93
Royal Irish Regiment, x–xi
Royal Ulster Constabulary (RUC), xiii

RPD. *See* recognition-primed model of decision-making
RPM. *See* recognition planning model of military decision-making
rumination, 160–63, 164–65, 172–73
   defined, 160–61
   following negative outcome, 163
sacred values, 92–93, 94–100, 119, 189
   defined, 92–93
   examples of adhering to, 94–99
   morally injurious experiences and, 169–70
sacrifice, 87–88, 89–90, 99–100
Saddam Hussein, 106
SAFE-T model, 17, 34–35, 50–52, 53, 59, 76, 77, 101, 120–21, 122, 138, 177–78, 185. *See also* plan execution; plan formulation; situational awareness; team learning
   benefits of, 34
   decision inertia and, 37, 42, 51
   decision-making impediments identified by, 42–49
   development of, 35
   military decision-making process and, 34, 35, 36, 37, 49–52
   utility of, 49–50
satisficing, 9–10, 158
Scott, Capt. (interview subject), 93–97, 130–31, 180–81
secondary cognitive appraisal, 155
secular values, 93, 94, 96–97, 119
sense-making, 53–54
   explained, 54–56
   hard decisions and, 7, 8–9
September 11 attacks, 165, 166
SERE training, 154
7/7 London bombings, 7–8
sexual violence and PTSD, 168
Sharm El Sheik bombings, 7
shell shock, 164
shoot/don't shoot decisions, 10, 11
shuras, 124n1
Siddle, Bruce, 138–39
signal detection theory, 162–63
situational awareness (SA), 24, 35, 49–50, 53–74, 187
   accountability and, 45–46
   defined, 35, 55–56
   error in, 64–73, 76
   hard decisions and, 5–6, 7–8
   levels of, 55–56, 59, 60, 61–62, 63, 64–65, 73–75
   OODA loop and, 49–50
   overview, 53–54
   sense-making and (*see* sense-making)
   uncertainty and, 47–48
   walk-in scenario, 56–60
sleep deprivation, 150–51
SOF. *See* Special Operations Forces
Soldier and Small Unit Performance Optimization (S2PO) strategy, 185–86
soldiers
   choice and, 84–85, 177
   civilian culture compared to culture of, 183–84
   civilians compared with on personality, 180–83
   definition used in text, xv$n$1
   learning from, 175–89
   the "subtle art of not giving a f*ck" and, 179–80
SOP. *See* standard operating procedure
Special Air Service, ix
Special Operations Forces (SOF), 102, 103
   interoperability and, 124, 127, 128, 130
   plan formulation and, 79–82
standard operating procedure (SOP), 23, 84, 185–86
startle response, 164–65, 166–67
status quo, 38
storytelling, 24–25, 53–54, 83
stress, 171–73. *See also* post-traumatic stress disorder
   in extremis, 153–56
   hard decisions and, 7–9
   individual differences in response, 154–56
   process-driven model of, 155
*Subtle Art of Not Giving a F*ck, The* (Manson), 179
sunk cost fallacy, 78–79
Survival, Evasion, Resistance, and Escape (SERE) training, 154
swift trust, 116–17

Synthetic Training Environment, 187
Syria, xvii–xviii, 10–11, 39
System 1, 9–10
System 2, 9–10
systemic stressors, 172

taboo trade-off, 93, 96, 97–98
Taliban, 15–16, 94, 96, 97–98
task ambiguity, 5, 8–9
task conflict, 106–7, 111–12, 118
   relationship conflict and, 112, 119
   trust and, 117
team learning (TL), 35, 120–37
   components of, 120–21
   defined, 35
   interoperability and (see interoperability)
tempo, 26–27, 30
10,000 Volts (10kV) reporting system, 6–8
Territorial Army, ix
"There Is Nothing Like a Soldier" (poem), 180
threat (stressor as), 155
threat rigidity, 135
threshold of acceptance, 10
time constraints, 178
   military decision-making process and, 19, 20
   naturalistic decision-making and, 24–25
   recognition planning model and, 26, 27
trade-offs, 87, 88, 99–100
   routine, 93
   taboo, 93, 96, 97–98
   tragic, 93, 97–98
traditional decision-making, 20–21
Trafalgar, battle of, xii
tragic trade-off, 93, 97–98
training. *See also* expertise and experience
   SERE, 154
   situation awareness and, 61–64
trait anxiety, 154
trimodal theory of decision-making, 77–78
trust, 177–78
   category-based, 117
   defined, 113
   interoperability and, 122–23, 126, 127–28
   in military teams, 113–18
   person-based, 115
   plan execution and, 101, 112–18
   role-based, 117
   swift, 116–17
tsunami, Boxing Day. *See* Boxing Day tsunami
tunnel scenario, 39–40
tunnel vision, 151–52
twin studies, 165
Type A personalities, 180–81

Ulster Defence Regiment, ix
uncertainty, 5, 178, 185–86, 187
   defined, 5
   endogenous, 47, 48–49, 121–22, 177–78
   exogenous (*see* exogenous uncertainty)
   hard decisions and, 5, 7–9
   military decision-making process and, 22–23
   plan formulation and, 47–48, 82
   preference, 85–86
   in the real world, 22–23
   SAFE-T model on, 47–49
United Kingdom Special Forces (UKSF), ix
utilitarian behavior, 182–83
utility theory, 21

value (hardness factor 6), 3, 4
value congruence, 105–6, 119
values, 177, 178–79, 189
   of armed forces *vs.* civilians, 183–84
   defined, 92–93, 105
   interoperability and, 128
   intra-team conflict in, 105–7
   plan execution and, 105–7, 119
   sacred (*see* sacred values)
   secular, 93, 94, 96–97, 119
Vietnam War, 164
vigilance tasks, 149–50

walk-in scenario, 56–60
war crimes, ix
war-gaming, 18–19, 27, 28
Watts, Alan, 179

weapons of mass destruction, 106
Webb, Maj. R. (interview subject), 56–60, 61, 77–78, 97–98, 161–63
West, Capt. (interview subject), 86, 98–99, 140–46, 148–49, 151–52, 153, 155–58, 163, 173–74

women, PTSD in, 168
working memory, 45, 74–75, 147–48
World War I, 73–74
World War II, 73–74

zone of indifference, 10, 158

www.ingramcontent.com/pod-product-compliance
Lightning Source LLC
LaVergne TN
LVHW011008250326
834688LV00004B/139